MOON MEN RETURN

MOON MEN RETURN

USS *Hornet* and the Recovery of the
Apollo 11 Astronauts

Scott W. Carmichael

Naval Institute Press
Annapolis, Maryland

Naval Institute Press
291 Wood Road
Annapolis, MD 21402

Library of Congress Cataloging-in-Publication Data

Carmichael, Scott W.
  Moon men return / Scott Carmichael.
     p. cm.
  Includes bibliographical references and index
  ISBN 978-1-59114-110-5 (alk. paper)
     1. Apollo 11 (Spacecraft)—History. 2. Hornet (Aircraft carrier : CVS-12)—History.
3. Space rescue operations—History—20th century. I. Title.
  TL789.8.U6A52433 2010
  629.45′4—dc22   2010009800

17 16 15 14 13 12 11 10   9 8 7 6 5 4 3 2
First printing

Contents

Preface

ON 5 June 1969, one of America's most storied and decorated aircraft carriers, USS *Hornet* (CVS-12), was selected to serve as primary recovery ship for the flight of Apollo 11. Her mission: Transport NASA and Navy recovery specialists to the designated splashdown site in the Pacific Ocean; locate the Apollo 11 command module, *Columbia,* after she re-entered earth's atmosphere and descended by parachute to the surface of the sea; stabilize and secure the spacecraft after splashdown in the ocean; recover the crew of American astronauts from shark-infested waters and transport them safely to USS *Hornet*; and finally, retrieve the command module itself, with its precious cargo of moon rocks, from the water.

The U.S. Navy had successfully executed the final search and rescue phase of manned spacecraft missions on many occasions since the first American astronaut was launched into a suborbital flight in May 1961. But this particular operation, the recovery of Apollo 11's command module after it returned to earth, was special, and undeniably so, for the flight of Apollo 11 represented the culmination of a national goal set by the late, beloved president of the United States, John F. Kennedy, to land a man on the moon before the end of the decade, and to bring him safely home to earth.

The final phase of the flight of Apollo 11 was a world event, and a very public one. Much of the world witnessed the return of Neil A. Armstrong, Col. Edwin "Buzz" Aldrin Jr., and Col. Michael Collins as it unfolded during live radio and TV broadcasts transmitted by major networks to every corner of the globe. American TV networks devoted more than thirty hours of broadcast time to continuous coverage of the flight prior to splashdown in the Pacific, and an estimated 500 million people remained glued to their TV sets around the world to catch a glimpse of men who actually walked on the moon. Hundreds of millions more listened to live radio coverage of the event as it was broadcast from makeshift studios located on USS *Hornet*'s flight deck. Approximately one quarter of all humankind remained awake and active, just to witness their return. A newly elected president of the United States, Richard Millhouse Nixon, flew halfway around the world from the White House, to land on *Hornet* as it steamed in the Pacific Ocean and welcome home America's newest heroes during a televised live ceremony at sea. The president's chief of staff, the American secretary of state, the administrator of NASA, and the commanders in chief of all American military forces in the Pacific (CINCPAC)—whose principal responsibility at the time was the active prosecution of an unpopular war in Vietnam—were all present for this most momentous of events: The return of the first men to walk on the moon.

Procedures to accomplish the recovery mission were developed by NASA during the early years of the manned spacecraft program, and were perfected during repeated recovery efforts throughout the 1960s. The agency initially considered the employment of the Soviet model of ending manned missions by landing their modules on solid ground—possibly by gliding the spacecraft to earth on fixed wings, or by paragliding them to relatively soft landings in open fields, or even by a cushioned descent onto solid soil by parachute, as the Russians continue to practice to this day. But in the end, the practicality of landing in the world's oceans won the day. Three quarters of the earth's surface is covered by water, and the relative ease of hitting water versus land during any given attempt made water landings an obvious choice. Besides, the United States of America employed a blue-water Navy at the time, which was capable of operating freely across the world's oceans. The final phase of each spaceflight, then, was a splashdown at sea.

Initially NASA considered the purchase of its own ships and the employment of a specially trained crew to conduct splashdown recoveries. The projected operating costs, however, were prohibitive, and a proposal for NASA to develop and deploy its own sea-based recovery forces was quickly abandoned. The agency turned, instead, to the Department of Defense (DoD) for support of its recovery operations—an assignment that the DoD was eager to take on. Projected to span at least a decade, the manned spaceflight program was the type of highly visible operation DoD embraced and a welcome alternative to the negative publicity endured by the department as a result of the war in Vietnam.

DoD created a multilayered command structure staffed by professionals who were thoroughly familiar with NASA's recovery mission and dedicated to maintaining an effective working relationship with NASA. The creation of a bureaucracy within the DoD to handle this mission ensured a consistently high level of support for NASA. At the top was the DoD manager for Manned Space Flight Support Operations, whose office was located within the Pentagon, in Washington, D.C. The functions of that office were largely administrative in nature: to address such issues as budget, policy, senior-level liaison with Congress and NASA, and to exercise overall responsibility for success of the DoD effort. Beneath the seat of government level, the manager for Manned Space Flight Support Operations staffed a liaison office within NASA's Manned Spacecraft Center in Houston, Texas, to coordinate DoD actions directly with NASA's Landing and Recovery Division. That NASA element, comprising approximately 130 NASA and DoD personnel, served as the nucleus for landing and recovery operations. It developed the procedures for recovery operations that were then executed by DoD elements at sea. Members of NASA's Landing and Recovery Division traveled throughout the world to train military elements in manned spacecraft recovery and retrieval procedures, and then participated in actual recovery and retrieval operations as advisers to military commanders on the scene. They were an integral part of the team: NASA provided expertise in recovery operations, while the American military and the DoD provided skilled and experienced assets at sea to execute NASA's recovery procedures.

The splashdowns themselves were initially envisioned to occur in both the Atlantic and the Pacific oceans. Separate military commands were created to handle operations in either theater. Task Force 130 (TF-130) was established to handle splashdowns in the Pacific, and Task Force 140 (TF-140) assumed responsibility in the Atlantic. Skeletal staffs at each command developed and administered training programs for military units selected to participate in recovery operations and also commanded U.S. military recovery forces "chopped" to the task forces for actual recovery operations.

At the operational level were any number of ships, aircraft, and special military units selected to execute recovery operations. Ships were plentiful, of course. Virtually any oceangoing vessel in the U.S. Navy's inventory could be equipped with special NASA equipment to locate a spacecraft and retrieve it from the water. But the ideal platform for such missions was an aircraft carrier or similarly large vessel that could handle both helicopter and fixed-wing aircraft to support search and rescue operations, and the more pedestrian task of retrieving or hoisting a spacecraft from the water once it was found. The Navy had only a handful of such vessels in the inventory, and none could be dedicated to solely support NASA's requirement. So a compromise was reached: Prior to the launch of each NASA spacecraft, the Navy selected a ship, usually a carrier or an LPH, to serve as primary recovery ship to support NASA's recovery operation.

Notification of such selections was commonly made no more than a few months in advance of a spaceflight—relatively short notice for such an operation. Once a selection was made, the designated PRS would interrupt its regularly planned schedule of operations and a team from NASA's Landing and Recovery Division, along with members of either TF-130 or TF-140, would board the PRS and prepare her crew for the recovery and retrieval mission.

To execute the most critical "hands-on" activity during recovery operations—the rescue phase—the Department of Defense provided its own teams of specialists. They were the first to make direct contact with the astronauts after splashdown; their mission was to find the spacecraft quickly within the vast expanse of open ocean and to rescue them in the event an emergency egress from the command module was necessary. NASA recovery operations specialist John C. Stonesifer, a veteran of more than twenty on-site recovery operations during the 1960s, characterized manned spacecraft recovery operations as search and rescue missions.

The goal of such operations was quite simply to find them, and then to save them. NASA was able to calculate with a remarkable degree of accuracy the general area on the sea in which a manned spacecraft might splash down. But any number of events could, and occasionally did, throw a spacecraft off target or cause the spacecraft to sink shortly after splashdown. Astronauts could find themselves alone, at night, hundreds of miles from their intended splashdown target, without communications, in a sinking spacecraft within shark-infested waters. Their very lives depended, then, on the ability of NASA and the military to quickly locate their spacecraft, get "hands on" the craft as it floated upon the water to stabilize it, and finally remove the astronauts from their vessel before it sank beneath them.

Aircraft carriers and other primary recovery ships steamed to the general area where NASA spacecraft were expected to splash down. But the critical functions of locating the spacecraft—getting eyes on target— and then placing hands on the craft to rescue astronauts in the event of emergency, required the employment of skilled and experienced men of resolve and determination who could be counted on to get the job done efficiently, regardless of circumstances.

The Department of Defense employed a number of specially trained units to execute the rescue phase of spacecraft recovery operations. Those units maintained a high state of training and readiness to perform their critical duties. Thus, where primary recovery ships might be selected at the last minute before launch of a particular spacecraft, elite military units that executed the rescue phase of recovery operations lived and breathed their roles daily.

Among those units was the United States Air Force's Aerospace Rescue and Recovery Service (ARRS), which deployed HC-54 and HC-130 turboprop aircraft to support search operations in the event a spacecraft either overshot or undershot its primary splashdown area. Extra fuel tanks enabled these specially configured aircraft to stay aloft for up to twenty-four hours and search large areas of ocean, as necessary. On board the aircraft, specially trained para-rescuemen—the storied U.S. Air Force PJs—were prepared to parachute into the sea after a spacecraft was located, to render assistance. Always stationed aloft during recovery operations, Air Force PJs were, on more than one occasion, deployed to rescue downed astronauts.

Operating from on board primary recovery ships were specially trained helicopter squadrons which were employed to locate NASA spacecraft as they descended by parachute to the surface of the sea. Spacecraft typically splashed down a good distance from primary recovery ships—ten miles or more—and, as they were generally "over the horizon" when they hit the water, the spacecraft could not be seen from recovery ships as they descended to the surface. Helicopters launched by primary recovery ships operated at a distance of ten to fifteen miles from the ships, located NASA spacecraft by homing in on emergency transmitters, placed "eyes on target" as the spacecraft descended toward splashdown, and then hovered nearby the floating vessels to assist with the effort to stabilize the spacecraft in the water, retrieve the astronauts, and then transport them to the recovery ships. The primary recovery ships themselves, were often miles away from the action but steaming in the general direction of the expected splashdown point, while helicopter crews and specialty swimmers ensured that the spacecraft was located and the astronauts were safe.

Finally, in the water, elite swimmers placed hands on the spacecraft and ensured astronaut safety until hovering helicopters could hoist the astronauts aboard for transport to the primary recovery ships. This was arguably the most crucial of tasks performed during a manned spacecraft recovery operation. In the event of emergency—if the spacecraft began to sink or if American astronauts found themselves in the water and threatened by sharks, as happened—teams of specially trained Navy swimmers joined the astronauts to render assistance. They were Navy commandos, drawn from the Navy's Underwater Demolition Teams (UDT) and arguably the toughest-minded men in the American military. Only the finest were chosen for the recovery of Apollo 11.

Acronyms

ADDRS	aircraft deployed drift reduction system
AN/SPS-30	long-range 3D height-finding radar system
ARRS	United States Air Force's Aerospace Rescue and Recovery Service
ARS	Air Rescue Service
ASW	antisubmarine warfare
ATS-1	Goddard Space Flight Center satellite system
B&A	boat and aircraft crane
BIG suits	biological isolation garments
CATCC	Carrier Air Traffic Control Center
CIC	combat information center
CINCPAC	commander in chief of U.S. military forces throughout the Pacific theater
CINPACFLT	commander in chief of the Pacific Fleet
CM	command module
COD	carrier onboard delivery aircraft (Northrop Grumman, manufacturer)
COMNAVAIRPAC	Commander Naval Air Pacific
DRT	dead-reckoning tracer
EM	enlisted men's club
GFCS	Gun Fire Control System
GUIDO	guidance officer
HS-4	Helicopter Anti-submarine Squadron 4
ICBC	Interagency Committee on Back Contamination
IFF	Identification Friend or Foe system

IMU	inertial measurement unit
INSURV	Board of Inspection and Survey
LORAN	Long-Range Navigation
LPH	Amphibious Assault Ship (Helicopter)
LTA	lighter-than-air
MARDET	Marine Detachment
1MC	main circuit; one-way public address system used on Navy vessels
MQF	mobile quarantine facilities
MTB	motor torpedo boat
NAS	Naval Air Station
NASA	National Aeronautics and Space Administration
NRO	National Reconnaissance Office
OI Division	within aircraft carrier Operations Department, responsible for tracking all surface and air contacts
PRS	primary recovery ship
SARAH	Search and Rescue and Homing system (World War II–era search and rescue system)
SDV	SEAL delivery vehicles
SIMEX	simulated exercise
SPS-30	radar antennae; air search radar system with high altitude capability
STAR	surface to air recovery
TACAN	tactical airborne navigation system
TACSAT	tactical communications satellite
TDY	temporary duty
UDT	underwater demolition team
UDTRA	underwater demolition training
UNIVAC	mainframe computer developed by Sperry-Rand Corporation
UNREP	underway replenishment operations at sea

1

USS *Hornet* (CVS-12)

U.S. Navy commander Chris W. Lamb served as the executive officer (XO) on board USS *Hornet* (CVS-12) when she was selected to serve as primary recovery ship (PRS) for Apollo 11. He attributes her selection for that assignment to chance—*Hornet* just happened to be the right type of vessel for the assignment; she just happened to be in the right place, as she was home ported on the West Coast of the United States; and she was available at the time a PRS was needed. She was lucky. But during a July 2007 interview, Commander Lamb, who retired as a Navy captain living within a stone's throw of the U.S. Naval Academy in Annapolis, Maryland, conceded that the selection of USS *Hornet* for the assignment seemed fitting, in that her record of performance in the service of her country, to that date, had been noteworthy. *Hornet* had earned the honor.

The Navy could not have selected a more storied or decorated ship to serve as the PRS for the recovery of Apollo 11. Launched in 1943 at the height of America's war in the Pacific and named after the sunken American aircraft carrier named *Hornet*, she was the eighth such ship to bear that name in the Navy's history. After her launch, *Hornet* joined the fray in the Pacific theater to reclaim territory seized by Japanese forces. She operated in forward areas of the Pacific combat zone for 16 continuous months. And she was lucky. *Hornet* came under attack by hostile forces fifty-nine times, but was never hit. She destroyed 1,410 Japanese aircraft during that period, a total exceeded only by her sister carrier, USS *Essex*. And she destroyed or damaged nearly 1.3 million tons of enemy shipping. Ten of her pilots achieved the status of "Ace in a Day"—accounting for the destruction of at least 5 enemy aircraft in a single day. Of the forty-two pilots assigned to her Hellcat fighter wing, thirty became aces. They recorded the shootdown of a record 72 enemy aircraft in a single day, and recorded the shootdown of a record 255 enemy aircraft in a single, very busy month. *Hornet* supported nearly every amphibious landing conducted by American military forces in the Pacific theater after March 1944, including operations in the Philippines, Guam, Iwo Jima, and Okinawa. She scored the critical first hits in sinking the super battleship *Yamato*. *Hornet*'s sting truly became legendary. Appropriately, near the end of the war *Hornet* launched the first strikes against Tokyo since the Doolittle Raid. She received nine battle stars for her service in World War II, and was awarded the Presidential Unit Citation for her World War II operations.

USS *Hornet* later performed service during the Korean and Vietnam conflicts. Her mission evolved over time, and when selected to serve as the PRS for Apollo 11, *Hornet* had been reconfigured to specialize in the conduct of antisubmarine warfare. In that capacity, *Hornet* no longer launched fighter or fighter-bomber aircraft from her decks, but rather transported squadrons of Sikorsky manufactured SH-3 helicopters and E1B early warning aircraft, which protected America's 7th Fleet then operating in the Tonkin Gulf off the coasts of Vietnam and the People's Republic of China, against the threat of attack by hostile submarines. As she neared the end of her life cycle, having accumulated twenty-six years of service in support of U.S. military operations since her launch, the promise of *Hornet*'s creed seemed to assure that the final phase of Apollo 11's historic flight would continue "A Heritage of Excellence," which defined the storied career of the carrier.

From 30 September 1968 until 12 May 1969, USS *Hornet* conducted operations in the Western Pacific under the command of Cpt. Jackson A. Stockton, USN. It was the ship's final WestPac cruise, an eight-month-long ordeal of continuous operations at sea which included a wartime stint in the South China Sea and the Gulf of Tonkin. Her final port of call in Yokosuka, Japan, on 1 May 1969 was a welcome relief. *Hornet* headed south and then east, to cross the Pacific Ocean en route to her home port in Long Beach, California.

As she departed for home, a newcomer boarded the ship. He was Cpt. Carl J. Seiberlich, a twenty-five-year veteran of the U.S. Navy and the man who was selected to replace Captain Stockton as commanding officer (CO) of *Hornet* when she arrived in Long Beach. Ordinarily a newly appointed CO would have waited for *Hornet* to return to Long Beach before reporting aboard. But Seiberlich wanted to get a feel for the ship and to take his measure of the crew before taking command of both. He should have been pleased by what he found. The crew was a cohesive and disciplined unit. They were a team. The men exuded a sense of pride in their work and displayed an esprit de corps which reflected the personable and easygoing style of their CO.

The ship returned to home port in Long Beach, California, on 12 May 12 and entered a routine period of transition. Maintenance commenced, and a good portion of the ship's crew took leave, departed for training, or transferred to new assignments. Former USS *Hornet* XO, Cpt. Chris Lamb, recalled that the routine turnover of the ship's crew posed a particular challenge to readiness on this occasion. Ordinarily *Hornet* could have expected to remain in port for many months before returning to sea. During that lengthy time in port, new men could settle into their assignments on USS *Hornet*, learn their way about the ship, and develop a feel for her routine and for their new shipmates. The passage of time would make up for the loss through normal attrition of the many experienced officers and enlisted crewmen who left USS *Hornet* after her return to Long Beach in May 1969. But *Hornet*'s time in port on this occasion was unusually short. She would have to accomplish her mission with many new men on board and with a new CO, as well.

Eleven days after reaching Long Beach, Captain Seiberlich took command of the *Hornet*. He was born on 4 July 1921 in Jenkintown, Pennsylvania, which was located less than fourteen miles from the site of the Liberty Bell in downtown Philadelphia. As a boy, Seiberlich joined the Sea Scouts and earned the Eagle Scout badge. Seiberlich received a degree from the University of Pennsylvania before going on to the U.S. Merchant Marine Academy at Kings Point, New York; he graduated in 1943. During World War II, Carl Seiberlich served in both the Atlantic and Pacific theaters of war. His skills as a navigator were honed at that time, during an assignment to USS *Mayo* (DD-422). It was a skill which would come in very handy during USS *Hornet*'s assignment to locate and recover the Apollo 11 command module.

Following the war, Seiberlich served on escort carriers in lighter-than-air (LTA) airships, or blimps, and in 1948 he made the first night landing on a carrier while at the controls of an LTA aircraft. He was then assigned by the Navy to develop the use of variable depth towed sonar from blimps for antisubmarine warfare (ASW), and in 1952, at age thirty-one, he received the Harmon International Aviation Trophy from President Harry S. Truman. By the time he assumed command of the *Hornet* on 23 May 1969, the ship's new captain had served as the operations officer and executive officer of USS *Valley Forge* (CVS-45), as the commanding officer of Air Anti-Submarine Squadron 26 operating from USS *Randolph* (CVS-15), and as navigator on USS *Intrepid* (CVS-11). Just before assuming his duties as the CO of USS *Hornet* in May 1969, Captain Seiberlich served as the CO of USS *Salamonie* (AO-26). He was just forty-seven years old and had served a total of twenty-six years with the Merchant Marine and Navy by the time he took command of one of the U.S. Navy's most formidable and storied warships. Over the course of his career he became the only U.S. naval aviator in history qualified to land three types of aircraft on aircraft carriers: fixed wing, helicopters, and blimps. His experience, then, with the operation of U.S. Navy aircraft carriers, and particularly with the operation of aircraft carriers that specialized in ASW warfare, such as USS *Hornet* (CVS-12), was unparalleled. Commanding the *Hornet* while she floated on the precipice of history marked the pinnacle of his career.

The splashdown of Apollo 11 was just two months away, and given the status of both the ship and her crew, the new CO of USS *Hornet* had his work cut out for him. According to officers who served under his command, Captain Seiberlich's watchwords were "preparation" and "safety." His penchant for paying attention to detail was tested shortly after assuming his duties as CO of USS *Hornet*, when the Navy's Board of Inspection and Survey (INSURV) conducted an inspection of the ship. The board's mission was to ensure that every U.S. Navy ship was fit and ready to accomplish its mission. Rear Adm. John D. Bulkeley served as

INSURV president at the time, and he personally conducted the inspection of USS *Hornet* (CVS-12).

Like Seiberlich, Bulkeley had amassed an impressive service record by 1969. A Medal of Honor recipient for heroic actions under fire while commanding a motor torpedo boat (MTB) squadron in the Philippines after the invasion by Japanese forces at the beginning of World War II, he was famously aggressive and remarkably successful in battle. He personally commanded the MTB boat that evacuated Gen. Douglas MacArthur, his immediate family members, and staff from Corregidor to Australia after the surrender of American forces to the Japanese. He later commanded the PT boats that cleared the way for the Normandy invasion. Following that action, Bulkeley personally selected then-Ens. John Fitzgerald Kennedy for command of Navy PT boats in the Pacific. After Kennedy was elected president of the United States in 1960, Kennedy hosted a luncheon for Bulkeley and his spouse at the White House, where they swapped war stories for an hour. Kennedy demonstrated his faith in Bulkeley when he selected the decorated officer to command Guantanamo Navy Base in Cuba and to face down Cuban dictator Fidel Castro after the Cuban Missile Crisis in October 1962. The Navy always seemed to give Bulkeley the toughest assignments, and he always produced results. It seemed fitting, then, for Bulkeley to be chosen to inspect the Navy ship that was tasked to complete President Kennedy's mandate to bring humankind's first moon-walking astronauts safely back to earth.

An INSURV inspection conducted by Rear Admiral Bulkeley was not a cakewalk. It was a serious business, from the top of the mast to the bottom of the keel. The previous sixteen ships to undergo an INSURV inspection by Rear Admiral Bulkeley had failed their inspections. USS *Hornet* was at a disadvantage in any event, of course, given her deteriorated condition after the eight-month-long WestPac cruise and the Apollo 11 recovery. But fortune and luck smiled on *Hornet* in the form of Cpt. Roy Goldman, a member of Rear Admiral Bulkeley's inspection team who happened to be a Naval Academy classmate of USS *Hornet*'s XO, Captain Lamb.

Naval Academy "ring knockers" tend to take care of one another. That is especially true of classmates. Goldman took Lamb aside and confided that the sixteen previous INSURV inspection failures were caused more by neglect of the admiral than by neglect of the vessels themselves. The COs of those vessels had tended to remain aloof from the admiral during INSURV inspections. The COs properly stayed out of the way of the inspectors, of course, but by removing themselves from the process and maintaining a distance from the admiral himself, the COs unintentionally communicated to Rear Admiral Bulkeley that they were possibly unconcerned about the inspection results. Rear Admiral Bulkeley taught each of them a lesson by "failing" their ships. Goldman recommended that the officers of *Hornet* take a different approach in their dealings with Rear Admiral Bulkeley by paying extraordinary attention to the admiral during the inspection and by communicating by their actions that they took the inspection seriously.

It worked. Captain Seiberlich ordered his department heads to personally escort inspection team members and to ensure that their senior NCOs accompanied the teams, clipboards in hand, to take copious notes during the inspection. The division chiefs organized, in advance, teams of men whose mission was to correct deficiencies on the spot, if possible, when deficiencies were cited by members of the INSURV inspection team, and to certainly correct such deficiencies no later than overnight. Meanwhile Captain Seiberlich accompanied the admiral personally.

The inspection lasted a full week. Ordinarily, a ship would be given several months to correct deficiencies discovered during an INSURV inspection. But *Hornet* did not wait to address them. Teams of *Hornet* crewmen swarmed over the ship at night to correct deficiencies cited by the INSURV inspection teams during the day. INSURV inspectors were surprised to learn that 85 percent of the deficiencies they cited during any given day were corrected by the time they returned to resume their inspection the next morning. The only major deficiency that remained uncorrected was the application of a coat of paint to the ship's hull, a task which could not be completed by crews at night. *Hornet* passed the inspection with flying colors, and Rear Admiral Bulkeley issued a personal commendation to Captain Seiberlich for his effort to ensure that USS *Hornet* was truly shipshape. Lt. Cdr. Robert P. Schmidt, engineering officer, also received the admiral's commendation.

Captain Seiberlich's approach to the INSURV inspection reflected an astute appreciation for the need to pay attention to detail, and a savvy approach to dealing with superiors. It also sent a signal to the crew of *Hornet:* Their new CO meant to get the job done; to get it done right; and to always ensure that his ship would be ready to accomplish her mission. Lt. (jg) Richard F. Powers III served as a young officer under Captain Seiberlich's command on board USS *Hornet.* He stated that Captain Seiberlich's unmistakable message to the crew during the INSURV inspection was that perfection was the only acceptable outcome for any USS *Hornet* endeavor while the ship operated under his command.

On 5 June 1969, less than six weeks before the launch of Apollo 11 on her mission to the moon, *Hornet* was formally nominated to serve as the PRS for the Apollo 11 recovery. Captain Seiberlich was consequently designated the commander of the Primary Landing Area Recovery Group. The ship was still moored on the starboard side to pier 2, berth 24 in her home port, the U.S. Naval Shipyard in Long Beach, California.

Many of her new crewmen had reported aboard. One of the latter was Ens. Thomas M. Meisenhelder, a Naval Reserve officer who reported for duty on 3 June, just two days before USS *Hornet* was nominated to serve as PRS for Apollo 11. It was Meisenhelder's first Navy assignment, and he was designated to serve on board *Hornet* as the OS division officer, in charge of the ship's sonar department. He would also serve as a watch officer within the ship's combat information center (CIC). Upon reflection, some forty years after completion of the Apollo 11 mission, Ensign Meisenhelder recalled that he was fully aware of the historic significance of the flight of Apollo 11 at the time he reported for duty on board USS *Hornet,* and that he was indeed a fortunate guy to have served during such a historic evolution. He told his parents to save everything they could find regarding the flight of Apollo 11.

The Apollo missions got off to an ominous start as tragedy almost derailed the program before a rocket left the ground. During a prelaunch exercise, a spark ignited a flash fire within the oxygen-enriched atmosphere of the Apollo 1 command module while the rocket remained affixed to its launchpad, and rescuers were unable to open the hatch in time to save our men. Astronauts Virgil I. "Gus" Grissom, Ed White, and Roger B. Chaffee were trapped inside the spacecraft and died from asphyxiation and exposure to extreme heat.

The danger only increased after the pre-launch phase. Indeed, during the early stages of the American manned spacecraft program, just getting a rocket off the ground without exploding in mid-flight or spiraling out of control seemed a challenge. Failures during liftoff seemed more common than successes. But none of our astronauts was lost during the liftoff phase of those early flights, and NASA seemed to have gotten its act completely together until the tragic 1986 launch of the *Challenger* space shuttle reminded us that even the launch phase of space travel can be dangerous when a single part fails.

Once in space, things can go wrong, too. A short circuit sparked an explosion within the service module during the flight of Apollo 13, which led to the abort of that mission and the near loss of her crew. Space is almost unforgiving. A mistake or a malfunction can lead to death.

No phase of space travel, however, has endangered the lives of astronauts more than the return to earth's atmosphere and the landing. The tragic return of the space shuttle *Columbia* in 2003 may be cited more often than others as exemplifying the dangers faced by astronauts when they return to earth. But *Columbia* was just one of many disasters and near disasters to occur during the final phase of spaceflight. A number of Soviet cosmonauts reportedly died when life support systems or parachute deployment systems failed, causing the astronauts to either asphyxiate in space or to die from trauma after plummeting from the heavens at terminal velocity to crash-land on earth.

Prior to *Columbia*, America suffered her own near misses and near disasters. Two early incidents highlighted this reality for NASA. In July 1961, Gus Grissom piloted the Mercury spacecraft, *Liberty Bell 7*, during a suborbital flight, to a splashdown site in the Atlantic Ocean. Some believe that Grissom prematurely completed a planned protocol to jettison the hatch of the spacecraft before rescue helicopters and swimmers were prepared for his exit, and that an inadvertent and premature activation of a detonation sequence "blew the hatch" and enabled seawater to flood the spacecraft. Grissom managed to swim free and was rescued. But efforts to recover the foundering spacecraft failed, and it sank to the bottom of the ocean. Eleven months later on 24 May 1962, Scott Carpenter piloted the Mercury spacecraft *Aurora 7* to a splashdown site after completing three orbits of the earth. But his spacecraft overshot the splashdown target area by 250 miles, leaving Carpenter stranded in the open ocean until USAF para-rescue swimmers A1C John F. Heitsch and Sgt. Ray McClure parachuted from their Air Force ARS SC-54 aircraft to rescue him.

Four years went by before another close call nearly claimed the lives of two astronauts. In March 1966, astronauts Neil A. Armstrong and David R. Scott were on board the Gemini spacecraft Gemini 8, when several malfunctions led to a decision to bring the spacecraft back to earth several days prior to the scheduled end of the mission. The spacecraft overshot its intended splashdown target area by more than two hundred miles, leaving Armstrong and Scott stranded in the Pacific Ocean, some one thousand miles south of Yokosuka, Japan, until a three-man USAF rescue team parachuted into the ocean to secure the spacecraft and assist the astronauts until a Navy ship arrived on the scene, some three hours later.

Finally, as the Apollo program entered its late stage in July 1971, one of three parachutes designed to slow the descent of the Apollo 15 command module toward the surface failed when caustic emissions vented from the spacecraft burned through a riser, leaving the 12,000-pound spacecraft suspended by only two fully deployed parachutes. The spacecraft descended rapidly toward a rough landing. David R. Scott, a veteran of the Gemini 8 spaceflight, and commander of the Apollo 15 mission, thus survived a second mishap in flight.

Fourteen minutes typically pass from the time a returning spacecraft enters the upper reaches of the earth's atmosphere until it touches down on the earth's surface. Under nominal conditions, recovery forces tracked the return of spacecraft using NASA ground-based radar stations and, in the immediate vicinity of splashdown zones, radar systems operated by pre-positioned Navy ships. When Apollo command modules descended to an elevation of ten thousand feet their main parachutes deployed, and an emergency radio beacon activated to signal their location to recovery forces below. USAF recovery aircraft positioned more than one hundred miles to the north and south of the primary splashdown zone employed direction-finding equipment to locate the descending spacecraft.

Down below, in closer proximity to the predicted splashdown point, Navy helicopters launched by the PRS employed a World War II–era search and rescue system, the SARAH Search and Rescue and Homing system, to home in on the spacecraft's recovery beacon. Astronauts inside the spacecraft established radio communications with recovery forces and provided information concerning their status and location. And during hours of darkness, a flashing light mounted on the spacecraft aided helicopter pilots in placing "eyes on target" to find the spacecraft, preferably before it splashed down, into the sea. Finally, Navy swimmers jumped from those helicopters to render hands-on assistance to stabilize the craft and to egress the astronauts to safety.

The idea was to locate the spacecraft quickly, preferably before splashdown, to ensure the safety of the astronauts. That was not always easily accomplished. Earlier flights demonstrated that conditions during the return phase of spaceflights were not always "nominal." Systems malfunctions on board an Apollo command module could cause the spacecraft to either undershoot or overshoot the primary splashdown point by hundreds of miles. The structural integrity of a command module could be compromised by a hard landing, or even by rough seas, causing it to take on water and possibly to founder and sink. Astronauts could be injured during re-entry.

To complicate matters further, Apollo command modules had an unfortunate propensity to turn upside down shortly after splashdown. While in that configuration, which NASA labeled "Stable 2," the emergency radio beacon, the flashing light, and the astronauts' radio antennae were underwater and inoperable. In short, the systems which recovery forces relied on to locate the spacecraft were not available while the spacecraft was upside down in the water. And the astronauts could not easily egress their craft, underwater, without assistance.

Apollo command modules remained stable while floating upside down, and they did continue to float, rather than sink. The spacecraft was equipped with airbags which righted the craft within five to eight minutes. But in the event the latter system failed, or if a rough sea state continued to push the spacecraft over into an upside-down configuration, rescue forces would find it difficult to locate the spacecraft at all in the open expanse of ocean. It was a particularly acute situation during hours of darkness, when many returning Apollo command modules, including Apollo 11, splashed down. So if a spacecraft either undershot or overshot its splashdown point by hundreds of miles during hours of darkness, and an Apollo command module overturned, as they often did, the astronauts would truly be lost at sea.

When the Navy sent a team to accomplish the recovery of Apollo 11, it sent its best team. They sent as their tactical commander,

Carl J. Seiberlich, because he was a detail-oriented man who understood the value of preparation and practice, and because he demonstrated during previous assignments a remarkable ability to get even the most difficult job done right. They sent USS *Hornet* (CVS-12), a decorated and storied vessel whose crew understood the value and need for teamwork as a key element for success. And they sent members of elite units whose specialized experience and training ensured that they would indeed place hands on *Columbia* and her crew, and bring them safely to earth. The Navy sent a team who got it right.

2

Apollo 11: The BIG One

AMERICA'S manned space flight program spanned almost a decade before the July 1969 flight of Apollo 11, and consisted of twenty manned flights conducted throughout the Mercury, Gemini, and early Apollo programs. But the flight of Apollo 11 constituted the ultimate objective for the program, within the context of the national goal set by President Kennedy on 25 May 1961. Everything done before that flight was preparatory for it. The lives that were lost, the lessons learned, the experience gathered, the procedures developed, and the materials and equipment tested throughout the early stages of the space program served to advance the spaceflight program toward the final goal of placing a man on the moon.

Public support for the program was initially based on a combination of fear and wonder. As the Cold War gained momentum, America imagined itself in a race to the moon with the Soviet Union. Rivalry between the two nations was further fueled by the successful launch in 1957 of *Sputnik*, the perceived "missile gap" that President Kennedy had made a focus of his campaign, and the Cuban Missile Crisis in October 1962. The moon was merely another piece in the geopolitical chess game between the two Cold War superpowers. Each country expected to demonstrate its superiority over the other by getting to the moon first.

At the same time, the very idea of launching men into space atop powerful rockets, landing them on the surface of the moon, and then returning them safely to earth seemed the stuff of science fiction. So the attempt to convert that dream into a reality sparked the public's imagination. Support for the effort among the American people was driven not only by fear that the Soviet Union might otherwise surpass the United States technologically and thereafter dominate the country politically, but also by a sense of wonder at the audacity of the attempt.

Against the backdrop of the civil rights movement and the Vietnam War, the flight of Apollo 11 offered a brief moment of unity to a divided nation. It was a mind-boggling attempt to accomplish for all people the remarkable feat of breaking free from the gravitational confines of our planet to walk on another celestial body. People, black and white alike, both peaceniks and hawks, paused for a moment within the daily turmoil of their lives to focus their collective attention in awe on three men as they risked their lives to accomplish the seemingly impossible. For all those grand and sweeping reasons, the flight of Apollo 11 was different from all such space shots before it.

But there were practical differences, as well, not the least of which was the fact that Apollo 11 expected to return to earth with samples of the moon. Moon rocks and moon dust. And possibly, moon germs. The fear that materials from outer space might return to contaminate the earth was indeed as old as Buck Rogers from the 1930s and perhaps older than that. But the real impetus for such concern was the October 1957 flight of the Soviet satellite *Sputnik*, the very first manmade object to circle the earth. In 1958 the chairman of the medical genetics department at the University of Wisconsin, Josh Lederberg, first publically voiced concern about the possibility that objects returning from space might carry with them foreign materials which could contaminate the earth. His thoughts were quickly adopted by the National Academy of Sciences, and in short order the International Council of Scientific Unions formed an ad hoc Committee on Contamination by Extraterrestrial Exploration to develop a "code of conduct" which would permit the exploration of space but at the same time prevent contamination by materials brought back from space. The U.S. National Academy of Sciences' Space Science Board, which served as an advisory group for NASA, referred the issue to NASA, with recommendations, and thus was born NASA's determination to deal seriously with the issue of back contamination by so-called moon germs.

In 1966 NASA's director and the surgeon general established the Interagency Committee on Back Contamination (ICBC) to ensure that NASA properly addressed the issue. The ICBC was authorized to review and approve plans and procedures to prevent back contamination. Its members were drawn from the Public Health Service, the Department of Agriculture, the Department of the Interior, NASA, and the National Academy of Sciences. But it was not until the flight of Apollo 11, and the very real possibility that NASA would actually transport lunar materials back to earth, that the prospect of otherworldly contamination of the entire planet might be in the offing. Back contamination in the form of "moon germs" then took on a life of its own, and many people developed a genuine concern for the future welfare of Mother Earth.

Given the potential for an emotional response to the threat, possibly in the form of widespread resistance throughout the world to NASA's moon shot, NASA was forced to additionally deal with back contamination as a public relations issue. Members of the ICBC met with NASA engineers to brainstorm some solutions to the problem. It was a meeting of very serious-minded theoretical scientists and a group of equally serious but practically oriented NASA engineers. Unfortunately, the two groups could barely relate to one another.

The ideal solution proffered by a member of the advisory committee would be for NASA to parachute the returning *Columbia* command module directly on the deck of the aircraft carrier USS *Hornet* (CVS-12) and to instantly seal the module in a plastic box of some kind. The NASA engineers assumed that he was not serious. But he was. The scientist had never been to sea. And he assumed that the achievement of pinpoint accuracy on a wave-tossed and therefore undulating flight deck of a rescue ship, which appeared to be the size of a postage stamp when viewed from outer space, was entirely possible.

A compromise was reached. NASA would develop materials and procedures to contain and quarantine returning moon rocks and whatever "germs" they might contain and to minimize the possibility of contamination. Biological isolation garments, or BIG suits, were created for the astronauts and for anyone and everyone who came into direct contact with the returning astronauts before they could be isolated and housed in a quarantine facility. The suits included masks which allowed the astronauts to inhale, but exhaled air was captured and filtered to avoid external contamination. Upon their return, the astronauts, their command module (CM), and everything that they touched prior to their entry would go into a quarantine facility on the big day where it would be sterilized. Betadine and a form of common household bleach made from sodium hypochlorite served as the sterilization agents.

Upon recovery from their CM after splashdown and their transport to USS *Hornet* (CVS-12), the astronauts would be housed in a specially constructed quarantine facility, a trailer, until that trailer could be offloaded and flown to Houston, where the astronauts would be placed into a larger quarantine facility until a minimum twenty-one days of quarantine could be completed. No prior spaceflight adhered to such procedures. It gave the appearance that Apollo 11 was very special indeed and that it was a potentially more dangerous undertaking, for all of humanity, than any prior flight.

The crew chosen for the flight of Apollo 11, who submitted to the wearing of BIG suits and an involuntary quarantine to protect the world from moon germ contamination after their return, were instant celebrities. All crewmembers of the flight of Apollo 11 were born in 1930. In their late thirties, they were a remarkably serious, quiet, focused, dedicated, and unassuming group of professionals whose purpose was quite simply to accomplish yet another mission. They were private people. But the relative anonymity in which they dwelled before the flight would be shattered thereafter by a blinding light of public scrutiny during a worldwide media frenzy, the likes of which had never before been launched on their native planet.

Neil Alden Armstrong, a civilian, was selected to serve as mission commander. Armstrong had demonstrated during training and previous space flights a remarkable ability to deal with stressful situations in a calm and detached manner. In a situation where men were about to embark on a mission not previously attempted by humans, and where the unexpected could be expected to occur, NASA needed a cool operator like Neil Armstrong on the team. He would be the first to step on the moon. And his name would likely become the answer to trivia questions posed throughout many generations with respect to space travel. Armstrong had served as a naval aviator and flew seventy-eight combat missions in Korea. He had also flown as a test pilot and once flew the X-15 at a speed of four thousand miles per hour into the outer edges of the atmosphere. He was among the second group of candidates to be selected to serve as an astronaut, and he commanded the Gemini 8 mission. At the time of his flight to the moon, Armstrong was being paid a salary of $30,000 per year.

Col. Edwin E. "Buzz" Aldrin Jr. was a graduate of West Point, and also a Korean War veteran who flew sixty-six combat missions during that conflict. He obtained a PhD in astronautics at the Massachusetts Institute of Technology (MIT), where he studied techniques to accomplish the rendezvous and docking of spacecraft while in orbit around the earth. Aldrin was among the third group of candidates selected for astronaut training, and he flew the final Gemini mission. Aldrin would follow Armstrong out the hatch after landing on the moon and would become the second human being to walk on another celestial body. Aldrin's pay and allowances as a full colonel at the time of his flight totaled little more than $18,000 per year.

Lt. Col. Michael Collins was also a graduate of West Point. He joined the U.S. Air Force, became a pilot, and accumulated a total four thousand hours flying experimental jets while assigned to Edwards Air Force Base in California—the program site for the book and movie *The Right Stuff*. Collins served with Aldrin as a member of the third group of candidates selected for astronaut training, and he flew in Gemini 10. Collins would serve as the CM pilot for Apollo 11 and remain aloft in that spacecraft while Armstrong and Aldrin walked on the moon. As compensation for his service during the flight of Apollo 11, Collins received approximately $17,000 per year in pay and allowances.

The increased level of public interest also changed the routine on board the Navy's primary recovery ship (PRS) for the recovery of Apollo 11. Previously, space flights had attracted only a small pool of reporters who accompanied the PRS during recovery and retrieval operations at sea; media interest in the flight of Apollo 11 practically overwhelmed the crew of USS *Hornet* with their demands for access and information.

Recognizing the significance of the upcoming spaceflight, ABC television signed an exclusive contract to film the recovery of Apollo 11 from on board USS *Hornet*, and to distribute footage to other networks and stations so that coverage of the proceedings could be shared with an expected worldwide TV audience of 500 million viewers, then the most widely broadcast TV event in history. This made sense not only from the corporate perspective of ABC, but from the logistical perspective of NASA, DoD, and the Navy as well. Space was precious on the primary recovery ship and limiting the television broadcast to only one of the three national networks at the time was a prudent decision on the part of mission organizers.

The media frenzy, however, was not limited to television. Besides the cadre of print journalists observing events, General Electric and the Mutual Broadcasting System prepared for live radio broadcasts to an estimated 300 million listeners around the world. Four nations refused to allow such broadcasts: Vietnam, North Korea, Albania, and the People's Republic of China. Voice of America prepared to broadcast live from the deck of USS *Hornet* in Russian to the people of the Soviet Union, who had great interest in America's attempt to land a man on the moon. Given the competitive, if not hostile, nature of relations between the United States and the Soviet Union at the time—the very height of the Cold War—Soviet willingness to cooperate with American media efforts was perhaps remarkable. As a result of these efforts, it is estimated that approximately one quarter of the earth's total population of 3.6 billion people in 1969 witnessed the moon landing in some manner.

USS *Hornet* augmented its small public affairs staff with temporarily assigned personnel to support a large body of pool reporters, numbering approximately 120, who reported on board *Hornet* during a stopover in Hawaii, just to cover the recovery effort. Their stories were filed and transmitted on a daily basis to newspapers and periodicals around the world. It was a zoo. Men and their equipment were everywhere on the ship. But it was expected. And the entire entourage was organized and handled quite professionally by a staff of Navy men whose everyday mission was to launch helicopters in search of enemy submarines.

The flight of Apollo 11 was such a historic event that even stamp collectors jumped into the mix with demands for "philatelic mail" that celebrated the date on which astronauts who walked on the moon returned safely to earth and were successfully recovered from the ocean. Such requests, numbering 250,000, were received from all over the world. The processing of these requests required USS *Hornet*'s mail room to affix a special date stamp to envelopes for return, as a souvenir, to the requester. That sounds easy. But just imagine the effort to stamp 250,000 such envelopes, all on the same day—24 July 1969—the date on which the CM for Apollo 11 splashed down in the Pacific Ocean. A volunteer team of twenty-five men from USS *Hornet*, under the direct command of CW1 John C. Varley, processed those requests that day. Captain Seiberlich personally handled VIP requests. Thus, the Apollo 11 recovery operation was far from a routine matter for the men charged with responsibility to recover the astronauts after splashdown in the Pacific Ocean and to retrieve the CM and its cargo of moon rocks. The pressure was on. Their actions were under the microscope of worldwide attention and were recorded for posterity. Their success or failure was very public.

The officer who stood most openly in the public eye, of course, was Carl Seiberlich, who assumed command of USS *Hornet* just weeks before the recovery took place. The least of Seiberlich's worries was embodied in Rear Adm. Donald C. "Red Dog" Davis, the commander of Manned Spaceflight Recovery Forces, Pacific, the Hawaii-based military element that assumed overall responsibility for military forces involved in NASA recovery operations in the Pacific. Anytime a flag-ranked officer boards a Navy ship for a visit, tension among the ship's officers and crew raises a notch. It is important to make—and equally important to leave—a good impression on the brass. Davis was an old Navy test pilot, and he was comfortable on aircraft carriers. His stay would be brief, just a couple of nights, and it is likely that Captain Seiberlich viewed the prospect of Rear Admiral Davis's visit as relatively innocuous, as long as everything sailed smoothly.

Of greater concern was the planned visit of Adm. John S. McCain Jr., the commander in chief of U.S. military forces throughout the Pacific theater (CINCPAC). Insofar as USS *Hornet* was homeported in Long Beach, California, the ship was part of Admiral McCain's command. Admiral McCain wore four stars on his shoulder boards. He had been in the Navy essentially forever—since he was a teenager—and had risen steadily through the ranks until he achieved near-legendary status. His nickname among Naval Academy graduates, who knew him more by reputation than by interaction, was "the lawnmower." If Admiral McCain did not like you or your positions or opinions on a topic, he would mow you down. And that would be the end of you. Admiral McCain was a feisty man, short in stature and the son of a Navy admiral. His son, also named John, was a Navy pilot imprisoned in North Vietnam at the time of the recovery of Apollo 11. (The son of CINCPAC, John McCain would, after his release from North Vietnam, enter the political arena, serve in the U.S. Senate, and run for president.) Admiral McCain planned to arrive on board USS *Hornet* the evening before the recovery. His visit would be brief. Captain Seiberlich could only hope that the recovery and all of the planned festivities which accompanied it would go off without a hitch.

But Captain Seiberlich's greatest concern, beyond the responsibility for his ship and crew and everything else that happened on board USS *Hornet* during recovery operations, was the planned visit to his ship by the president of the United States, his commander in chief, President Richard M. Nixon. President Nixon had been sworn in to his office exactly six months before Apollo 11's lunar module touched down on the surface of the moon. Not long. He had inherited the Apollo program from presidents Kennedy and Johnson and was poised to bathe in the glory of its crowning achievement. His visit would constitute the first time a president of the United States had personally observed a splashdown, and the fact that he was prepared to fly halfway around the world, from Washington, D.C., to the Pacific, to participate in the welcoming ceremony on board a Navy ship for the world's first men to walk on the moon said something about his determination to play the event for all it was worth politically.

Nixon's visit, though, however brief, tended to complicate matters. The first consideration, of course, was security, and *Hornet* had to make special accommodations to meet Secret Service requirements for Nixon's brief stay on the ship. The Secret Service advance team set up a command in the flag plot to coordinate actions related to presidential security for President Nixon's three-hour visit. And then there was the back contamination issue. It was not commonly known at the time, but a decision had already been made to quarantine the entire ship for up to twenty-one days if procedures to prevent the spread of moon germs were somehow breached during the recovery process. That meant, at the very least, that the president of the United States could find himself stuck on a U.S. Navy ship at sea for twenty-one days if efforts to contain returning moon germs were deemed to be in question. Not good. The president's visit was a great honor for USS *Hornet* and her crew, of course, but it also added to the burden of a newly appointed commanding officer.

The presence of so many high-ranking officers and the nation's commander in chief underscored the significance of the mission. It was to be a historic moment, but one for which *Hornet* was not yet prepared in May 1969. Seiberlich had other more pressing concerns than entertaining the Navy's top brass and the president in the weeks leading up to the recovery mission.

3

Departure from Long Beach

Hornet was still moored in her home port of Long Beach, California, when she was selected on 5 June 1969 to serve as the primary recovery ship (PRS) for the recovery of Apollo 11. Also moored in Long Beach was USS *Princeton* (LPH-5), an amphibious assault carrier which had successfully completed her assignment as the PRS for Apollo 10. The overlap of these two ships at the U.S. Naval Station in Long Beach during the last few weeks in June 1969 proved fortuitous for the crew of *Hornet*.

The *Princeton* performed a flawless recovery of Apollo 10's command module (CM) in the Pacific Ocean. The module returned to earth during daylight hours on 26 May 1969, under clear and sunny skies. Calm seas awaited her as she splashed down within two miles of the ship. Virtually the entire crew of the *Princeton*, along with a complement of pool reporters from the major broadcast and print media, plus a team from NASA's Landing and Recovery Division on board, watched from the flight deck as the module slowly descended toward the surface of the water beneath three billowing parachutes to an uneventful splashdown, practically within a stone's throw of the ship. *Hornet* benefited from *Princeton's* experience.

Five days after her selection as the PRS for the Apollo 11 recovery, a special Apollo spare parts package for HS-4, the ASW helicopter squadron on the West Coast that specialized in Apollo recoveries, was transferred from the hangar bay of *Princeton* to *Hornet*. The Department of the Navy ensured that Navy assets involved in manned spaceflight recoveries were adequately supplied and supported. The Navy would not allow a helicopter involved in an Apollo recovery to be grounded for lack of maintenance or spare parts. So a designated "package" of spare parts was created for the Sea King helicopters to use to support Apollo recovery missions—gauges, filters, and everything that might require replacement on Sea King helicopters were packed onto pallets for shipment to whichever ship was selected to serve as PRS for an Apollo recovery. An inventory established that the package was 80 percent complete. Requisition orders were submitted on a priority basis for replacement parts to complete the package.

Two days later, on Thursday, 12 June 1969, a PRS planning conference was held aboard USS *Hornet*. For the first time, Sieberlich, Lamb, and other officers received copies of the Apollo Recovery Operational Procedures Manual, authored by NASA's Landing and Recovery Division. *Princeton*'s after-action report from the recovery of Apollo 10 was also provided to those in attendance. NASA's manual served as the very heart of *Hornet*'s planning for the recovery; *Princeton*'s report passed along lessons learned while following NASA's recovery procedures.

Soon after that initial meeting, NASA engineers and *Hornet* crewmen began preparing the ship in earnest for the recovery. A NASA team arrived from Houston to assist with the installation on USS *Hornet* of special recovery gear, some of which was truly on the cutting edge of technology. Its sophistication absolutely delighted the crew. ETR3 Dennis R. Moran was in charge of *Hornet*'s radar repeater crew which maintained twelve radar repeaters, or workstations, on board the ship. NASA wanted the repeaters to be in perfect alignment, naturally, to ensure the accuracy of bearings and ranges to targets—in this case, the Apollo command module as it entered earth's atmosphere and then descended by parachute toward the surface of the ocean. Given NASA's deep pockets, Moran and his pals were encouraged to replace most of the cathode ray tubes and resisters in *Hornet*'s radar equipment. From Moran's point of view, the coolest thing about NASA was their technology for the Identification Friend or Foe (IFF) system—because it was beyond state of the art. NASA had already entered the digital age of computers; their IFF system offered a near-instantaneous display of data, which was little short of miraculous to technicians in 1969.

IFF pods were installed on *Hornet* aircraft to emit coded signals which uniquely identified aircraft to radar operators on the ship—thus, helicopter #66 carried an IFF pod which emitted a signal that identified the helicopter to radar operators as helicopter #66, not some other aircraft. With IFF pods activated, radar operators could distinguish one aircraft blip on their scopes from others, and thereby enable air traffic controllers to "visualize" the relative positions of each aircraft vis-à-vis *Hornet*. But state-of-the-art technology employed by USS *Hornet* radar operators in 1969 displayed not numbers on a radar scope, but small blobs which, if expanded through the manipulation of control knobs mounted on radar repeaters or workstations, appeared to be bar codes—similar to the bar codes used to scan items in a grocery store today. If a radar operator wished to "decode" the bars in order to identify an aircraft, he would count the lines and spaces on each bar code to translate the "code" into a number, like "66."

The system was time-consuming and awkward. But the NASA retrieval team changed all that when they reported for duty on board *Hornet*. They attached to *Hornet*'s radar scopes a small box which automatically converted the IFF "bar codes" to digital displays. All an operator had to do in order to produce a digital display of the number was to place a cursor on any target depicted on the scope and push a button. It was "point-and-click" technology, similar to that found on any laptop computer today. Moran observed that, in 1969, when scientists were still using slide rules to perform calculus functions and handheld calculators were yet to be marketed to the masses, such technology was simply unheard of, and amazing. ETR3 Moran considered himself to be a real technology geek back then, someone who read all of the trade magazines. But he was absolutely blown away by what he characterized as NASA's new "voodoo technology."

Not everything that NASA technicians touched while on board *Hornet* turned to gold. Moran recalled that NASA technicians ran a test with the ship's SPS-30 radar antennae which went bust. The SPS-30 was an air search radar system with a high altitude capability. It was configured with manual controls which enabled an operator to point at a specific object, or to specific and narrow regions in the sky, rather than sweep the horizon in broad swaths. The SPS-30 was nearly perfect for NASA's requirement to locate and track an inbound Apollo command module as it screamed into the earth's atmosphere. The system's antenna was located at the base of the mast on the inboard side of the aircraft carrier's island. During a trial run, NASA technicians cranked the antenna downward so fast, its transmission jammed, rendering the entire system useless. Moran was already working fourteen-hour days as it was, and he really did not need the extra work, but he cleaned up NASA's mess and got the system up and running again.

Despite the occasional mishap, it was evident that NASA's team had developed some experience in preparing Navy ships for NASA recoveries. They arrived with checklists in hand, knew exactly what was needed, and worked cooperatively with members of the crew to get the job done. Some members of *Hornet*'s crew learned during the process, though, that even low-tech requirements to support NASA's recovery mission required the exercise of personal initiative and ingenuity.

One such requirement was for lighting. NASA technicians explained that the Apollo 11 command module was scheduled to splash down in the Pacific just before dawn, and that it would still be dark when the recovery was accomplished. The word "dark," in the open ocean, really means dark. Those involved in the hands-on aspects of the recovery and retrieval operation would not be able to see a thing unless sufficient artificial light were available. For UDT swimmers in the water, operating in close proximity to the spacecraft, the Navy mounted powerful spotlights on hovering helicopters to bathe the area around the floating CM in light. The swimmers would be able to see what they were doing. But after the astronauts were removed from their spacecraft and flown to *Hornet*, helicopters would remain in the immediate area of the floating CM only until *Hornet* made a close approach to the spacecraft to commence its retrieval operation—a time during which the CM would be hoisted by crane onto the ship. As the ship approached the module, recovery helicopters would clear the area and leave the module alone and unlighted. At that point, *Hornet* would have to be prepared to flood the area around the module with artificial light to support the retrieval operation.

NASA's team required only that *Hornet*'s twenty-four-inch carbon arc searchlight be in good working order. They turned to EM3 Michael P. Gasho and his colleague EM3 Michael Lee Laurence for assistance. Gasho and Laurence were USS *Hornet* electricians responsible for illumination of the ship's flight deck during flight operations. They typically worked twelve-hour shifts whenever *Hornet*-based helicopters operated from the carrier, day or night. Their duties required them to maintain the strobe lights, the landing lights, and every other kind of light which might be found above *Hornet*'s hangar deck. But neither of them had ever even heard that USS *Hornet* had available for their use a 24-inch carbon arc searchlight. The NASA technicians assured them that, indeed, all *Essex*-class aircraft carriers, like *Hornet*, were equipped with such a searchlight. It could be found high above the flight deck, on the "island," at the O-6 level. And, sure enough, that is where they found it. It was covered by an old tarp which appeared to have been secured sometime during World War II. Gasho and Laurence uncovered the searchlight, opened it, and examined its working components only to discover that it was hopelessly inoperable. If a maintenance record existed for that particular piece of equipment, they had no idea where it might be found. Not that it mattered. The searchlight was junk.

Gasho and Laurence were relatively junior petty officers. Had they consulted with superiors about the problem, they would have learned that the ship's requirement to support an Apollo recovery operation was of sufficient priority to ensure that a request for the requisition of a new searchlight would have been favorably received by the Navy, and a new searchlight would likely have been delivered by the Navy's supply system post haste. But Gasho and Laurence did not know any of that. They had no inkling that their requisition request would warrant priority handling. Their experience with the Navy supply system suggested, to the contrary, that the system would be unresponsive to any such request. In fact, they doubted that the system could produce an old 24-inch carbon arc searchlight at all. So they exercised some initiative, and in consonance with a time-honored tradition of U.S. Navy service that dated back to the very birth of the nation, they found a way to replace the inoperable 24-inch carbon arc searchlight on the O-6 level of USS *Hornet*'s island: They stole someone else's searchlight. More accurately, they conducted a midnight requisition—in broad daylight.

Gasho and Laurence no longer recall which Navy ship suffered the loss of its searchlight through their efforts. They do recall that it was another *Essex*-class aircraft carrier—and therefore identically configured to *Hornet*—that was in port in Long Beach, California, at the time. Thus, it was either the USS *Bennington* (CVS-20) or the Apollo-10 PRS, USS *Princeton* (LPH-5). Gasho and Laurence recall that they simply boarded the ship in broad daylight, as though they were crew members. They made their way to the O-6 level, found the searchlight, and examined its components to discover that it appeared to be in perfect working order. Unlike their own searchlight, this one had been well maintained. Not missing a beat, Gasho and Laurence unbolted the searchlight from its mounting and enlisted a couple of "yard apes" to assist them in securing it to a pallet. The pallet was lowered to the pier, and then transported by pickup truck to *Hornet*. After they mounted the new searchlight onto their own ship, Gasho and Laurence reversed course and bolted the inoperable searchlight onto the other ship. The switch was completed without discovery, and Gasho and Laurence kept that secret for thirty-nine years before finally revealing it, with some relish, to this author. They were home free, and NASA was satisfied.

While a handful of *Hornet*'s crew scrambled to prepare for the upcoming Apollo mission, life on board a U.S. Navy combat vessel continued unabated. On 13 June 1969, AA Michael A. Zamora, a member of the ship's 2nd Division, caught his left hand between a cleat and the rope while doubling up the standard mooring lines of the ship. USS *Hornet* was not a rowboat. When she shifted or rocked along the pier, some serious weight was brought to bear against her mooring lines which produced unimaginable pressure. Zamora suffered a traumatic amputation of his second and middle fingers and was transported to the naval hospital at Long Beach. The next day, MMFA William M. Van Hook reported to the naval station's dispensary with an injury, and before *Hornet* departed her home port, SA H. C. O'Connor of 1st Division suffered a laceration to the middle finger on his right hand, which was smashed between a float and the port side of the ship. Navy ships are dangerous environments in which to work, and injuries were common. *Hornet* would suffer her fair share of routine injuries throughout the operation to recover Apollo 11.

From 21 to 25 June 1969, *Hornet* on-loaded support equipment from ABC, General Electric, and the Mutual Broadcasting System. Three ABC

TV vans were placed in the forward end of hangar bay #2, on the same level and in the same general area as that designated for storage of two mobile quarantine facilities (MQFs) to house the three Apollo 11 astronauts in quarantine after splashdown. Lt. Henry Francis Dronzek, who served as the executive assistant to Captain Seiberlich during the operation to recover Apollo 11, interacted with ABC technicians who manned those vans, and recalled that they were the same vans used by ABC to televise professional and college football games during the 1960s. ABC had the contract to film an onboard welcoming ceremony which was planned for the astronauts shortly after their splashdown and recovery. One van housed equipment that received feeds from a number of fixed and mobile ABC cameras located throughout the ship, in order to dual record the event onto two identical two-inch magnetic reels. ABC intended to share that data with the other major TV networks and to retain a backup reel for posterity.

Other equipment followed ABC's. A General Electric (GE) Transatel terminal and associated units were loaded onto the flight deck in an area aft of the island but forward of the #3 elevator. That equipment supported the effort to broadcast the welcoming ceremony via satellite throughout the world. And a Mutual Broadcasting van, manned by technician James Joseph O'Connor, was also placed on *Hornet*'s flight deck, positioned on the tractor ramp on the starboard side of the island. Mutual Broadcasting's role was to broadcast radio coverage of the event to an estimated 300 million listeners all over the world.

Activities were taking place all over the ship. USS *Hornet* was forced to leave its helicopter squadrons and their support elements at home, to make room for equipment stored by NASA's onboard recovery team; by HS-4, the ASW helicopter squadron which supported Apollo recoveries on the West Coast; by UDT Detachment Apollo, the Navy combat swimmer detachment which supported Apollo recovery operations on the West Coast; and by the pool reporters and the various media representatives who expected to broadcast news of the recovery operation throughout the world. Only regularly assigned USS *Hornet* personnel would make the trip. The ship's roster dropped, therefore, from approximately 5,000 to 1,800 personnel. Storage space and berthing spaces were opened and available to temporarily house the visitors.

Lt. (jg) Peter D. Beaulieu, 1st Division chief, recalled that on Wednesday, 25 June 1969, NASA technicians arrived to install a special NASA winch onto *Hornet*'s boat and aircraft crane (B&A crane). The B&A crane would be the primary means of retrieval of the CM from the sea after splashdown. Its mission would be to hoist the 12,000-pound spacecraft vertically from the surface of the ocean onto the deck. From there, it would be towed to an elevator and taken down to the hangar deck for processing and storage. The winch had previously been installed on USS *Princeton*, which was then in port. So it was just a question of transferring the winch from the *Princeton* to *Hornet*, and then effecting its installation. The procedure required most of the day. Beaulieu was one of a handful of USS *Hornet* officers designated to supervise the retrieval operation.

The NASA team submitted to *Hornet* a detailed list of items and materials which NASA expected to provide to *Hornet* in support of the recovery and retrieval operation. They also submitted a list of items and materials which they expected *Hornet* to provide. And it was a detailed list. On Thursday, 26 June 1969, *Hornet* loaded the following items in order to fully comply with NASA's requirements: Two one-hundred-foot rolls of chicken wire, thirty iron posts, an equal number of five-gallon plastic buckets, and sixty bags of cement. It was a low-tech requirement. NASA anticipated a need to secure a perimeter around the MQFs and the CM once the astronauts were brought on board after splashdown, to prevent curious and gawking sailors and civilians alike from approaching the immediate areas in which the MQFs and CM were positioned on the hangar deck. The required materials would be used to construct a temporary fence. NASA was indeed a detail-oriented agency which attempted to think of everything.

Later that day, USS *Hornet* departed its home port, the U.S. Naval Station in Long Beach, California, to begin its journey to the Pacific Ocean for the recovery of Apollo 11's command module, *Columbia*, her crew, and her precious cargo of moon rocks. Her mission was historic, and one might imagine that the sole focus of her crew, at that moment, was on the need to prepare for the challenge ahead. But in fact, USS *Hornet* remained a working ship of the U.S. Navy, and her crew was therefore often required to accomplish a multitude of tasks in support of sometimes divergent requirements. They had to do it all. So before taking her leave from Long Beach, *Hornet* loaded into her hangar bay a number of Navy aircraft for transfer to San Diego. It was cargo, and *Hornet* received orders to haul it. But then, she was headed in that general direction, anyway.

USS *Hornet* did leave behind one piece of equipment in Long Beach that might have served her well during the days leading up to the recovery and certainly would have enabled *Hornet*'s navigator to avoid near disaster during the final phase of the recovery effort: a new satellite navigation system, just recently deployed to a handful of Navy ships. Launched during the early to mid-1960s, these satellites where designed to assist U.S. Navy vessels in their efforts to simply transit accurately from point A to point B on the globe. Shipboard equipment to exploit that capability was scarce. It was deployed on only a handful of Navy ships by the time Apollo 11 lifted free from her launchpad, and USS *Hornet* was not among them. Thus, on a mission to recover the first humans to walk on the moon, ship's navigators were forced to navigate the ocean by sextant and the stars.

Ordinarily this would have posed no problem. After all, navigation by sextant is what *Hornet*'s crew was used to. They were experienced and proficient, and they had alternate systems which were not available to navigators in the past. The Navy had for many years charted sections of the ocean floor by use of fathometers to measure the distance from the surface of the sea to the ocean floor. Seamounts, mountains which commonly rose to elevations thousands of feet above the sea floor, were discovered in this way, charted, and used by the U.S. Navy as navigational aids. *Hornet*'s navigational charts included many such seamounts. A more sophisticated system was the LORAN system of radio transmitters strategically placed in spots around the Pacific Ocean to broadcast homing signals to ships at sea. A navigator had only to lock onto several such signals to pinpoint his exact position on the globe. With funds made available to support NASA recovery operations, *Hornet* purchased an SPN-40 LORAN to extend the LORAN range.

Hornet, however, faced a number of challenges which were not easily overcome. The first challenge was a weather system in the Pacific Ocean which commonly caused the region near the primary splashdown target zone to be overcast. Overcast means clouds, and clouds tend to block one's view of the stars. If a weather system persisted, then *Hornet* would not be able obtain a fix on a sufficient number of stars to accurately fix her position on the globe. The second challenge stemmed from the circumstance that NASA wished to avoid: targeting her splashdowns in commercial shipping lanes. That certainly made sense. After all, NASA would not want a CM to suddenly drop from the sky onto or in the path of a cargo ship at sea. So NASA deliberately chose for its splashdown target areas, regions of the Pacific Ocean that were remote, far from shipping lanes. Those areas are so remote that the U.S. Navy had not yet charted the ocean floor with their fathometers to locate seamounts for use as navigational aids. *Hornet*'s charts were lacking. And, because the LORAN system was developed to assist commercial shipping, the system was less effective in remote regions of the Pacific Ocean than in those areas commonly used by cargo vessels.

Despite the potential difficulties, the ship's command element remained confident in their ability to navigate, and with good reason. The new commanding officer (CO), Captain Seiberlich, had served during a previous Navy assignment as navigator for a U.S. Navy warship—he was therefore experienced in navigating at sea. *Hornet*'s executive officer, Commander Lamb, served as the ship's navigator throughout her most recent WestPac cruise. Again, he was experienced. Additionally, and not least of the factors which weighed heavily in her favor, the ship retained an experienced quartermaster corps, a group of enlisted sailors who specialized in navigation. The ship's new navigator, Cdr. Robert A. Costigan, a Navy pilot who had just reported for duty on USS *Hornet*, had no previous experience at sea as a navigator. But he did purchase with Apollo-related funds a new 6×30 monocular sextant to augment the standard Navy sextant. The new sextant was easier to use to locate and identify stars during limited visibility—such as the commonly cloud-covered skies of the Pacific Ocean. *Hornet* seemed to be prepared.

By mid-morning on Friday, 27 June 1969, *Hornet* had made good progress on her run southward toward San Diego, California, for a brief stopover. The distance between Long Beach and San Diego was approximately 120 miles. Between 0900 and 1030, she landed a total fifteen aircraft on her flight deck, including a squadron of eight Sikorsky SH-3D Sea King helicopters from HS-4, the Black Knights, a unit based at the Naval Air Station Imperial Beach, California, near San Diego, which participated in the recoveries of Apollo missions 8, 10, 11, 12, and 13. She also accepted a total of four E-1B fixed-wing aircraft, which would directly support the recovery mission by providing communications relay and air boss command and control over the recovery site. And finally, three C-1A CODs, the Northrop Grumman–manufactured carrier onboard delivery aircraft, landed on *Hornet*'s flight deck. One of the CODs was regularly assigned to *Hornet*; the others were provided by VR-30, to provide daily and routine transport of materials and personnel to and from the aircraft carrier both before and after the recovery. The aircraft would remain on board *Hornet* throughout the recovery operation and until her return to Long Beach, California.

USS *Hornet* moored for just three hours at the Naval Air Station North Island, Coronado, California, on Friday, 27 June 1969, to off-load cargo, and to on-load five F-8 Corsair Crusader aircraft for transport to her next destination. By 1700 that day, she was under way en route to Pearl Harbor, Hawaii.

4

Transit to Pearl

AT 1700 on Friday, 27 June 1969, USS *Hornet* departed Naval Air Station North Island en route to Pearl Harbor, Hawaii. Three weeks had passed since she was selected to serve as the primary recovery ship (PRS) for the recovery of Apollo 11, and a small measure of progress had been made to prepare her crew for that assignment. Some of the equipment needed to accomplish her mission was on board, but not all of it. Her officers had a general understanding of the procedures they would follow during recovery and retrieval operations, but theirs was book knowledge only. Hands-on practice had not yet occurred and less than four weeks remained before Apollo 11 was scheduled to return to earth and splash down in the Pacific.

To prepare for the recovery, the first order of business was to create a team of men who would perform the retrieval operation. *Hornet*'s complement at that time totaled on the order of 1,800 men, but not everyone would be needed to man lines and operate machinery. Captain Seiberlich wanted to employ only the best. A memo circulated to division and department chiefs to nominate a handful of men from each element of the *Hornet* organization to staff the Hornet Module Retrieval Team. Only the most experienced, most competent, and most reliable men would qualify. A total of 90 would be needed. Cdr. Harley L. Stuntz III, the ship's weapons officer, was designated to lead the team.

Stuntz was a ring knocker, a 1955 graduate of the U.S. Naval Academy in Annapolis. As the ship's weapons officer, Stuntz already commanded most of the ship's divisions which would be principally involved in the retrieval effort. He was therefore well suited for the assignment as leader of the Hornet Module Retrieval Team. Stuntz was quick to praise the accomplishments of subordinates and to pass credit for his successes along to them.

Commander Stuntz's organization was divided into five teams. The first was a command module retrieval team, under the command of *Hornet's* first lieutenant, Lt. Cdr. Richard I. Knapp. Its mission was to capture the Apollo 11 command module, long after splashdown and after the astronauts were recovered by the UDT swimmers and transported to *Hornet* by helicopter. As the module bobbed about in the ocean, supported by a flotation collar installed by the UDT swimmers, and still accompanied by those swimmers, *Hornet* would maneuver alongside the module. A gunner's mate on board *Hornet* would then fire a "shot line" across the water to UDT swimmers standing on the command module (CM), to capture and secure the module before it drifted by. Members of the Hornet Module Retrieval Team would then hoist the CM onto the deck of *Hornet* by use of the ship's B&A crane, which was mounted on the starboard side of the ship.

BM2 Mickey N. Lowe was a member of the command module retrieval team. He recalled that, as the "go-to" guy for a senior chief boatswain's mate in the 3rd Division, Lowe was nominated to lead one small element of the Hornet Module Retrieval Team on deck. Lowe related that men selected for the team were drawn from throughout the ship and, for the most part, had not previously worked together. But each man was among the best. Each had demonstrated a positive mental attitude and an ability to work with others as part of a team to "get the job done" without offering excuses. And though many members of the team had not worked together before, quite a few of them at least knew one another from previous assignments. The gunner's mate who fired the shot line to UDT swimmers, for example, had served with Lowe on river patrol boats at the Naval Support Activity in Da Nang, Vietnam, just months before the Apollo 11 recovery. The gunner's mate's boat, #1477, was badly hit during an ambush, and Lowe's boat pulled up alongside to remove four wounded sailors for transport to a hospital. Lowe and the gunner's mate therefore had a shared experience during a previous assignment that served as a foundation for the positive work relationship they enjoyed during the recovery of Apollo 11.

The second team within Commander Stuntz's organization was the tracking and acquisition team, under the command of the ordnance officer. That team comprised fire control technicians and ordnance division officers. They manned the MK56 Gun Fire Control System (GFCS) forward, designated Radar 51, and the MK56 GFCS aft. They also manned the MK37 GFCS as a backup. Their team was responsible for acquiring the CM on their radar systems after splashdown, and for providing to the officer of the deck (OOD)—who was steering the ship—range and relative bearing data so that *Hornet* could vector toward the CM for retrieval. In practice, the MK56 GFCS radar did not lock onto the CM itself. It was unable to do so, as the CM bobbed up and down in the swells, and lost itself in the static. The fire control radar locked, instead, onto the helicopters which hovered above the CM during their recovery operation. Once locked onto those helicopters, *Hornet* could then plot its own course toward the CM for the later retrieval operation. Radar 51, which was physically located on the starboard side, forward of *Hornet*'s island, was expected to first acquire the helicopters and thereby locate the CM as it floated on the sea. Lt. (jg) Charles Stephen "Chuck" Rand served as the officer in charge (OIC) of Radar 51.

The third team within Commander Stuntz's organization was the security team, commanded by the commander of the USMC detachment on board *Hornet* (the MARDET), Cpt. Robert D. Caskey, USMC. Men assigned to this team were drawn from both the MARDET and the ship's master-at-arms force. The responsibility for this team was the security of the mobile quarantine facilities (MQFs), which were positioned in hangar bay #2; security for the CM, after it was brought on board and positioned in close proximity to one of the MQFs; and security support for the U.S. Secret Service team which would accompany President Nixon during his visit on board *Hornet*. Captain Caskey had earned a reputation as a resourceful officer. Lamb and Stuntz were confident that Captain Caskey would resolve all security-related issues which arose during the recovery operation.

The remaining two teams in Cdr. Stuntz's organization were the recording team. The recording team was charged with responsibility of collecting photographs and gathering notes which would be used to prepare a USS *Hornet* Apollo Cruise Report to document the event. The UDT assistance team was under the command of GO division officer (aviation ordnance), CW1 Albert L. Brown. That team's responsibility was the care and feeding of the men of UDT Detachment Apollo, which would join the ship after its arrival in Pearl Harbor.

Ordinarily Brown was in charge of the Red Shirts who could be seen on the flight deck of *Hornet*, handling munitions. His office was physically situated just beneath the flight deck, and above the hangar deck—"up in the struts," where noise and commotion were an everyday occurrence. Commander Stuntz had great confidence in Brown, who had earned a reputation for getting things done, never sitting on them, and always finding a solution to problems. Brown was a "go-to" guy. You could task him to perform the most difficult of assignments and walk away with full confidence that Brown would instantly respond to the tasking and get the job done right, the first time. He was the kind of man who made his boss look good.

Cdr. Harley Stuntz was in charge of the Hornet Module Retrieval Team, and answerable to the executive officer and commanding officer of the ship, for its success. But he selected good men to take care of the details and to get the job done right. As Harley explained most sincerely, nearly forty years after the recovery of Apollo 11: "I had good men under me who did all the work. Those guys did it. They deserve the credit, not me." If anything went wrong, however, it was Harley Stuntz who would find himself called on the carpet—the carpet in the XO's stateroom.

Lt. Henry Francis Dronzek, the executive assistant to Captain Seiberlich, recalled that the XO routinely hosted department head meetings in his stateroom every night at 1900. to gauge the progress of preparations, to identify additional issues to be addressed, and to assign tasks to specific officers for completion. The XO's job was to ensure that things got done on the ship. He paid attention to the most minute of details. Lamb possessed a likeable personality, ordinarily, but Lieutenant Dronzek recalled that the Apollo meetings were very intense, very thorough, and no-nonsense business affairs.

Sundays are usually a day of relative leisure, even on a working ship of the U.S. Navy, but Sunday, 29 June 1969, was anything but a day of leisure for the officers and senior enlisted men in charge of the Hornet Module Retrieval Team. They spent much of that day reviewing NASA's Apollo Recovery Operational Procedures Manual.

Miscellaneous gear provided by *Hornet* to support the recovery operation was gathered together and placed into a locker. The exceptions were 1,200 feet of 1-inch nylon line, two trash cans, one ice container, and a mess deck table. NASA's list was detailed. And everything on the list would be used.

On Monday, 30 June 1969, twenty-three bags of Redimix cement were mixed—with salt water, to avoid wasting freshwater. Thirty posts were then placed into thirty-five-gallon plastic buckets and set with cement. The security team would take possession of the posts after the cement had a chance to dry, and use them to cordon off the Apollo 11 command module, and the MQF which housed the astronauts after their splash-down and recovery.

The first meeting and general briefing of all Hornet Module Retrieval Team personnel was also held that day. The purpose of the briefing was to provide an overview of the operation for the majority of the team— those who had not had an opportunity to read and study the manual. It was a verbal briefing only, to generally acquaint the entire team with the concept of recovery operations. There were no walk-through exercises that day. No hands-on practice. Senior officers briefed the men on the sequence of events that they would follow leading up to the retrieval of the CM, and on the general safety requirements for retrieval operations. They displayed drawings which depicted the stations that would be occupied by specific teams of men during the retrieval operation, and which depicted equipment that would be used by the men during that operation. When the men were dismissed, a briefing was provided to the XO and the CO.

The next day, on Tuesday, 1 July 1969, the day before USS *Hornet* arrived at Pearl Harbor, Hawaii, Commander Stuntz, the weapons officer in charge of the Hornet Module Retrieval Team, reviewed with his officers and leading petty officers (LPOs) the NASA recovery procedures.

Hornet's preparations for the Apollo 11 recovery and retrieval operation prior to her arrival at Pearl Harbor were therefore largely academic in nature, intended only to generally acquaint *Hornet's* team with the sequence of events which the team would follow, and the specific procedures which the team would employ during the recovery and retrieval operation. Until they arrived in Pearl Harbor, they had no hands-on experience working with specialized NASA recovery and retrieval equipment. They were a team, but a team that lacked experience.

U.S. Navy aircraft carrier operations continued on *Hornet* as the ship approached Pearl Harbor, unabated by preparations which members of the Hornet Module Retrieval Team made for the recovery and retrieval of Apollo 11. Three helicopter pilots completed carrier qualifications that day. They were officially, and finally, fully capable of flying in the vicinity of and landing on an aircraft carrier that was operating at sea.

5

Pearl

USS *Hornet* arrived at the Naval Station Pearl Harbor at 1048 on Wednesday, 2 July 1969, and moored starboard side to Pier H berths 3 and 4 at Ford Island. She "chopped" to CTF-130, by which she administratively subordinated herself to the commander of Task Force 130, Rear Adm. "Red Dog" Davis, who wore two hats, serving also as the commander of the Hawaiian Sea Frontier. *Hornet* was officially on the rolls to accept orders from the command, which exercised overall responsibility for DoD support to NASA recovery operations in the Pacific.

No sooner had *Hornet* moored at the docks than a laundry list of recovery-related equipment began to be loaded on board. A spare parts box; a space recovery hook for the B&A crane, and *Hornet*'s "Tilley" crane, a mobile crane located on the carrier's flight deck to serve as a backup for the B&A crane in the event the latter failed or some unforeseen tactical situation made the Tilley a better choice for retrieval of the command module (CM); a CM cradle and tie-down ring; ratchet type tie-down rings and straps; line threaders; a 1,200-foot roll of .5-inch nylon line; a recovery sling assembly; an Apollo "workstand"; six training flotation collar boxes; two 50-ton shackles; and more. Fortunately, members of NASA's onboard recovery team were intimately familiar with every piece of specialized equipment that would be needed to accomplish the recovery and retrieval of the Apollo command module. Their role throughout the operation would be to both teach and advise. To the members of Commander Stuntz's Hornet Module Retrieval Team, much of the equipment was at least familiar, but some of it was downright foreign, if not otherworldly.

Arguably the most significant item to be loaded that day was boilerplate #1218. The boilerplate was a mock-up of the Apollo command module. Constructed to the same external dimensions and the approximate weight of a CM, it was a dummy module used during practice sessions. Finally, *Hornet* had available for hands-on practice a piece of equipment that resembled the spacecraft which they were scheduled to retrieve from the Pacific Ocean in just three weeks' time. The boilerplate came with its own dolly and transport dolly. It was pulled by a tractor, commonly called a "mule," into the hangar deck and positioned near elevator #3.

NASA's onboard recovery team included a number of specialists who performed liaison functions throughout the ship, to guide and advise Navy personnel on specific recovery- and retrieval-related activities. Every aspect of the operation was overseen by NASA specialists in that fashion. NASA liaison personnel did not perform the work, and they issued no orders or commands, but they did know what they were about and offered advice when needed. Early on, their advice and guidance were much in demand.

NASA's team additionally included technical specialists who provided hands-on support to install, operate, and/or monitor equipment dedicated to the recovery mission. One such technician was a technical representative for HS-4—an engineer—and HS-4 technicians installed SARAH equipment on seven HS-4 helicopters. Four of the receivers had to be properly aligned. SARAH receivers were employed by HS-4 helicopters to locate the CM as it dropped by parachute during an approximate five-minute descent from an altitude of approximately ten thousand feet to the surface of the sea. The module's emergency homing beacon automatically activated when the drogue chutes deployed, to broadcast an electronic signal for rescue aircraft to home in on.

Five minutes is not a lot of time in the open ocean, especially if rescue aircraft are not truly in the immediate vicinity. It is a race. As the CM descends, the line-of-sight distance over which the signal broadcasts decreases. And when the CM hits the water, the effective range of its homing signal is reduced to no more than five miles or so. Rescue aircraft, like HS-4's Sea King helicopters, must work feverishly to locate the CM—to actually get "eyeballs on target"—preferably before splashdown occurs and before the spacecraft loses its signal altogether after it rolls over into a Stable 2, upside-down, configuration in the water.

A SARAH package for Sea King helicopters included two receiving antennae, mounted like pointed fingers on each of two struts beneath the aircraft. The incoming signal from the CM's homing beacon was received by the antennae and routed to a workstation located in the passenger compartment of the aircraft. The combined signal was then displayed on a screen in the form of a spike, for a technician seated at the workstation. If the spike was in the middle of the screen, or exactly halfway across the screen, equidistant from both the left or right side of the screen, then the technician knew that the helicopter-mounted antennae were pointed directly at the homing signal. He cued the pilot by telling him to continue straight ahead on the course which he was currently traveling. If the spike displayed on the screen was way over to the right, then the technician could direct the pilot to bank farther to the right until the spike was once again dead center in the middle of the screen. In that fashion, the technician was able to vector his pilot on course to the CM. Given enough time on the correct course, the Sea King could even close enough to the CM to actually see it with the naked eye as it splashed down.

There were limitations, however, to the equipment. For example, SARAH equipment could not discern the distance between the Sea King helicopter and the CM. Thus, pilots did not really know how far away the CM might be from their current position. They only really knew whether the CM was dead ahead, or perhaps somewhere off to the left or to the right of their current course. Technicians could, with practice, sometimes gauge by signal strength alone whether their aircraft was near or far away from a CM. A very strong signal might indicate that the CM is very near indeed; while a very weak signal could indicate that the CM was quite some distance away from the helicopter. Variables such as signal strength and weather conditions could fool even the most experienced technicians, however, and that fact did come into play during the recovery of Apollo 11.

Alignment of the antennae, then, was critically important. If SARAH antennae were not properly aligned, the system provided incorrect data and the Sea King would miss its target entirely. The NASA technical representative for HS-4 spent much of the day, during that first day in port in Pearl Harbor, ensuring that the alignment was dead on.

On the following day, Thursday, 3 July 1969, *Hornet* officers met to address a number of administrative matters. Communications officers were briefed on procedures to be followed to ensure that NASA and Defense Department offices in Houston and Hawaii were linked to vessels involved in recovery and retrieval operations. The public affairs officers from *Hornet*, NASA, and CINCPAC met for briefings on the handling of the more than one hundred civilian reporters who were expected to board *Hornet* for the operation, and for worldwide radio and television coverage of the event. *Hornet* personnel were informed that they were expected to track and document recovery-related expenses incurred during the mission, principally in the form of purchases of materials and equipment procured specifically to support the operation, and that a summary report would be submitted by the Navy to NASA after the event for later reimbursement. No detail seemed too small to be addressed.

Thursday was a red-letter day for the Hornet Command Module Retrieval Team, because it marked the beginning of their formal training for the recovery and retrieval of Apollo 11. Crewmen assigned to the Hornet Command Module Retrieval Team were assembled for a formal briefing. Captain Seiberlich introduced the leaders of the thirty-five-member NASA recovery team which would be on board the ship throughout the recovery process: Dr. Donald E. Stullken, PhD, who regularly served as the chief of the Flight Operations Division at NASA's Manned Spacecraft Center in Houston, Texas, and Dr. Stullken's assistant, John C. Stonesifer, who was principally responsible for decontamination procedures to neutralize lunar pathogens.

It became clear during the lengthy and detailed briefing which followed that Dr. Stullken and John Stonesifer were serious about ensuring the safety of the astronauts. Dr. Donald E. Stullken was the inventor of the flotation collar which Navy combat swimmers installed around the Apollo command module after splashdown to stabilize the craft. His drive to create such a collar was a key to understanding the commitment to excellence which he demonstrated throughout the recovery effort, and which reflected NASA's determination to bring astronauts safely back to earth.

In 1958, the Army's Ballistic Missile Agency in Huntsville, Alabama, invited the Navy to participate in experiments with recoverable nose cones for Jupiter missiles. The Army launched an effort called Project Noah, to place a variety of life-forms into the nose cone of a Jupiter missile, including amoeba, bacteria, frogs, and mice, for experimental purposes. They also wished to place a monkey into space, in anticipation of an eventual manned spaceflight program, and they invited the Navy to design a life-support system for the animal, plus a telemetry package and bioinstrumentation to monitor the monkey's physiology during flight.

They chose as a test animal a South American squirrel monkey, and the Overhaul and Repair Department (O&R) of NAS Pensacola's fabric shop custom-built a capsule for the monkey within a Jupiter missile nose cone. Dr. Stullken worked on the life-support system, and developed a CO_2 scrubber, to remove CO_2 from air inside the capsule. When the buildup of dust on that scrubber proved problematic, Stullken demonstrated a remarkable degree of resourcefulness by applying an off-the-shelf material, then commonly used to stiffen women's petticoats, to contain the dust while allowing gases to flow freely through his filter. He built the contraption on his wife's sewing machine at home.

Army Missile 13, designated AM-13, was launched from Cape Canaveral on 13 December 1958, and splashed down near San Juan, Puerto Rico. A new agency at the time, NASA sent representatives, including Stullken, to observe the first ballistic animal flight. The flight was a success, and bioinstrumentation developed by the Navy clearly indicated that the South American squirrel monkey survived the launch, re-entry, and the impact of splashdown. But AM-13, launched on December 13, appeared to have been jinxed. A new bonding material used by the Army for its heat shield on the Jupiter nose cone failed; the nose cone developed a crack between the heat shield and the nose cone casing upon impact with the water; it leaked and sank. According to bioinstrumentation readings, the nose cone's passenger, the South American squirrel monkey, was alive and fully conscious as the nose cone sank to the bottom of the ocean.

Stullken was stung by the loss, however, and perhaps suffered pangs of guilt related to it. But he learned from it, as well. The loss made it clear that NASA needed to recover human astronauts quickly after splashdown, in the event an injury prevented them from managing their own escape from a sinking spacecraft. There was also an additional need for some kind of flotation device to maintain buoyancy during egress to prevent further disaster.

When Project Mercury was approved, NASA hired Dr. Stullken to conduct survival training for the astronauts. He recalled sharing a raft with John Glenn and Scott Carpenter during a survival exercise when a storm blew in and forced the trio to row for shore. Fortunately, the raft held together and kept them afloat until they reached land.

At the time, Stullken was also working with the U.S. Air Force on development of a twenty-man life raft. It occurred to him that, what NASA needed was a life raft–type device for its spacecraft. Stullken tasked the life raft shop of NAS Pensacola's O&R to design a flotation collar for the Mercury spacecraft, with a view toward providing both stabilization and flotation for the craft after splashdown. Thereafter, O&R developed, designed, and manufactured all flotation collars employed for Mercury, Gemini, and Apollo spacecraft.

The collar, then called a Stullken collar, was available but not used for Gus Grissom's ill-fated flight of *Liberty Bell 7* in July 1961, which sank. It was first used for Scott Carpenter's flight in May 1962 and kept his spacecraft afloat for two and a half hours until the primary recovery ship (PRS) arrived on the scene to retrieve it. The collar worked. During a later examination of Carpenter's collar at the O&R shop, someone removed a NASA patch from the collar and provided it to Dr. Stullken as a memento.

Stullken and Stonesifer participated in the recovery at sea of every manned NASA spacecraft throughout the approximately eight-year history of the program. They literally wrote the book on Apollo recovery procedures. Their objective on Thursday, 3 July 1969, was to describe for the ninety men of the Hornet Module Retrieval Team the sequence of events which normally occurred during an Apollo recovery and retrieval, and to explain to the team how and where they fit into that sequence. Thereafter, they intended to drive the Hornet Command Module Retrieval Team to practice to perfection. Neither astronauts nor spacecraft would be lost on their watch.

Members of Dr. Stullken's team were familiar with the special retrieval equipment used and with the specific procedures to be followed during Apollo command module retrieval, and all were experienced in Apollo retrieval operations. But NASA's role was advisory only. Members of the NASA team would instruct and advise, but they would not issue orders or directives. In order to ensure a clear chain of command, and to thereby avoid confusion and resultant errors, Navy officers would issue all commands. It was expected that, after a period of practice, Hornet Module Retrieval Team members would develop their own expertise in Apollo retrieval operations and would rely less on guidance from onboard NASA team members.

Procedures briefed by members of the NASA recovery team seemed simple and straightforward, in theory. But operations are often more difficult when put into practice at sea, where a pitching deck, bad weather, and lack of visibility during hours of darkness make life more complicated. There was nothing that Captain Seiberlich could do about the weather, sea states, or visibility during retrieval operations. But he made clear during the wrap-up of their initial training session that there was plenty that he could and would do about the development of expertise and teamwork. He concluded the training session by stating that he did not want any mistakes during the actual recovery. Training was the key. "You will retrieve this thing as you train." And he meant to train them until they executed their assignments with perfection.

Hornet was still in port when the retrieval team's first formal training session concluded. They could not, therefore, practice at-sea retrievals with the boilerplate. But practices for other aspects of the recovery and retrieval operation were possible. They walked through the motions of retrieval exercises while operating the motor whaleboat, the B&A crane, and the NS-50 Tilley crane which was positioned on the port side of the flight deck, just forward of elevator #2. The drills were merely walk-through exercises to acquaint teams with their assignments and with specialized NASA equipment, and to begin the process of ironing out kinks which always manifest themselves when a new team first tries to accomplish a mission.

Crews manning the ship's motor whaleboat, a twenty-six-foot powered water craft, were expected to serve as on-the-water assistants to the UDT swimmers and to the men on board *Hornet*, in whatever capacity they could. The "procedures manual" directed the launching of a whaleboat as the ship made its final approach to come alongside the CM during a retrieval operation. To provide the swimmers required extra equipment or to off-load unneeded equipment, members of the whaleboat crew would assist. If a shot line fired from the ship to swimmers on the CM fell short or fell off target, the whaleboat crew might retrieve the line and hand it to a swimmer. If equipment or other gear fell overboard, from either the CM or swimmer rafts or even from the ship itself, whaleboat crews might recover it before it drifted off. Whaleboat crews were essentially utility workers. They were on site to assist however they might.

One function performed by a member of the whaleboat crew was of particular interest to UDT swimmers working in the water around the CM: shark watch. Sharks normally follow Navy ships at sea. They are scavengers, and Navy ships routinely dump food waste off their fantail and directly into the ocean for disposal. As they steam their way through the water, Navy ships leave a wake of food particles which naturally attract the wolves of the sea. It was not uncommon for at least two sharks to be observed in the wake of *Hornet* during operations at sea. The splashing activity around a CM also attracted sharks, and on more than one occasion swimmers were forced to seek higher ground on a boilerplate or CM's flotation collar to avoid the swarms of curious and sometimes aggressive sharks which gathered round them. At least one member of the whaleboat crew was armed, but he was instructed to fire not directly at a shark, if firing at all were necessary, but rather simply to fire nearby. It was hoped that the sound and concussion produced by gunfire might scare the sharks away. Shooting a shark, of course, was forbidden unless a swimmer or astronaut or other human being was in immediate and direct danger of being bitten. Blood in the water could only make matters worse.

While members of the Hornet Module Retrieval Team exercised on deck and nearby within the port, their executive officer (XO), Commander Lamb, busied himself with a matter of executive importance: President Nixon's military aide, Lt. Cdr. Charles R. "Chuck" Larson, had arrived with a laundry list of presidential requirements documented in script on a yellow legal pad. Among them: a recommended menu for the presidential breakfast, which would be served to Nixon during the early morning hours on Splashdown day, while at sea on board *Hornet*. Lieutenant Commander Larson would eventually rise during a stellar military career to the position of CINCPAC, in charge of all U.S. military forces throughout the Pacific theater, and he would twice serve as superintendant of the U.S. Naval Academy in Annapolis, Maryland. But on 3 July 1969, Lieutenant Commander Larson's duty was to ensure that Richard Nixon was properly fed. The chief executive preferred Texas Ruby Red—pink—grapefruit for breakfast, and his choice of breakfast cereals was Wheaties, "the breakfast of champions."

Nixon required more than just a special breakfast menu. He required a leather couch and a set of matching leather chairs in the event he needed to rest during his brief three-to-four-hour stay on board the ship, during which he would observe recovery operations and then lead a televised welcoming ceremony for the astronauts after they arrived on board the ship. Under no circumstance could any sailor approach within twenty-five feet of the president, for security purposes. Such requests may sound mundane and laughingly minor to those unfamiliar with military culture, but for the XO they were of serious concern. Requests from the commander in chief were tantamount to orders, and some of them were nearly impossible to satisfy.

Lamb sent sailors scrambling all over the island in search of pink grapefruit, to no avail. None could be found. Nixon would have to settle for yellow grapefruit—selected specially for their freshness, firmness, and flavor—and Lamb could only hope that his failure to provide pink grapefruit for the president would go unnoticed or that it would be forgiven. And with no real time to requisition the furniture, Lamb led a midnight requisition at the officer's club at Ford Island the night before *Hornet* left Hawaii to recover Apollo 11. He and his men found the place locked, of course. They broke in, found the requisite leather couch and matching chairs, and loaded them into a Navy pickup truck. Lamb left a note to account for the theft, along with a promise to return the property after the recovery of Apollo 11.

The security requirement for the president, though, was a nonstarter. More than two thousand men would man *Hornet* during the recovery of the Apollo 11 astronauts. Many of them would be assigned to duties on the flight deck and the island structure of the aircraft carrier where the president could be expected to spend some time. The island, in particular, provided cramped quarters. And hundreds of crewmen would naturally be on hand to observe the welcoming ceremony for the astronauts in hangar bay #2. There was no way to operate a working vessel at sea, especially under these special circumstances, while requiring men to maintain a minimum distance of twenty-five feet from the president. Lamb scratched that requirement off the list, but promised to work cooperatively with the president's Secret Service detail to develop an alternate plan, a compromise, to meet their security requirements.

Lamb's efforts to satisfy seemingly picayune requirements for the president's visit only added to the burden which an XO ordinarily shoulders while running the day-to-day operations of a Navy warship, even one which is not at sea, but remains tied to the pier at Pearl Harbor. At the end of the day, Thursday, 3 July 1969, Lamb was informed that an enlisted member of his crew was arrested by local law enforcement authorities and turned over to the custody of *Hornet's* master at arms force. His offense: indecent exposure and association with a transvestite. It seemed that life in port could be as dangerous for a young sailor, though in different ways, perhaps, than at sea.

FRIDAY, 4 JULY 1969

It was Captain Seiberlich's birthday—the all-American boy who was born on the fourth of July near the nation's birthplace of Philadelphia, Pennsylvania. But the commanding officer of USS *Hornet* did not spend the day idly or in celebration.

At about 0200 that morning, an enlisted member of *Hornet's* crew awakened Cdr. Harley L. Stuntz from a sound sleep in his stateroom, located directly beneath one of *Hornet's* noisy catapults, to inform him that the captain required his presence. Stuntz was in charge of the Hornet Module Retrieval Team, and therefore responsible for virtually all aspects of *Hornet's* support for the Apollo 11 recovery and retrieval effort. And that included fresh paint.

Captain Seiberlich had discovered during the night, a small spot of rust on the underside of the Admiral's ladder—a metal stairway which would be used by VIP visitors to transit from one deck within the ship's island to another. It seemed unlikely that an admiral or any other human being on earth would bother to walk underneath that ladder and squint within the shadowed darkness to find a spot of rust, much less that a person of such stature would ever bother to mention it, in any event. But Captain Seiberlich was a detail-oriented skipper. He noticed. And he would have none of it. Stuntz ordered a subordinate to repair it immediately, at two o'clock in the morning. The damaged item could not wait for sunrise. And Seiberlich had made his point.

Lt. Henry Francis Dronzek served as Captain Seiberlich's administrative assistant and grew very close to him. He described Seiberlich as a very disciplined man and as an expert ship handler, who devoted twenty-four hours per day to the successful recovery of Apollo 11. He was very intense and extremely focused. Accuracy, safety, preparation, and attention to detail were his watchwords. He had no time for idle chatter.

XO Lamb recalled that Captain Seiberlich spent much of his time walking around the ship "playing what-if"—actively looking for problems and potential problems, and then formulating solutions in order to prevent those problems from ever seeing the light of day. Lamb kept his nose to the grindstone, as well, working to resolve day-to-day issues regarding the ship's operation. That devotion to duty left his commanding officer free to focus upon the bigger picture.

Lt. (jg) Peter D. Beaulieu documented in a government-issued memo book that, on this day, lines were painted on the deck from elevator #3 to the storage location in hangar bay #2 where the CM would be placed after its retrieval from the sea. A tractor, commonly called a "mule," would haul the CM on its dolly after retrieval and follow the freshly painted pathway to its final destination in the hangar bay, and thereby avoid any confusion or mistakes along the route. The effort stemmed from the old KISS axiom, which exhorted men to Keep It Simple Stupid. And the fresh paint also served as a visual reminder to the crew that the captain expected perfection throughout every phase of the operation. The message which they received: Attention to detail = No mistakes.

SATURDAY, 5 JULY 1969

Additional preparations for upcoming practice sessions were completed on Saturday. The tensile strength of the line on the B&A crane was tested to ensure that it could lift the 9,000-pound boilerplate mock CM without problem. The operational command module, *Columbia*, would weigh an additional three thousand pounds, and similar lines on the B&A crane would be tested prior to the actual retrieval on Splashdown day to ensure that they, too, were strong enough to handle the strain. Nothing was left to chance.

Liftoff of Apollo 11 was scheduled to occur in just eleven more days. The landing of Apollo 11's lunar module, *Eagle*, would occur on 20 July, and the return to earth and splashdown were scheduled to occur on 24 July, just nineteen days away. The Hornet Module Retrieval Team had yet to practice a single retrieval of the boilerplate, and members of UDT Detachment Apollo had yet to report for duty on board the ship.

SUNDAY, 6 JULY 1969

Bruce Erickson was a member of *Hornet's* #3 catapult crew, but was detailed to the mess decks during the operation to recover Apollo 11. He recalled that, as a young man, the brief stopover in Hawaii represented the first such visit to the exotic islands for him and other young members of the crew. They decided to sleep on the famous beaches of Hawaii at night, rather than return to their cramped berthing spaces on board the ship. But just as they settled in for a comfortable night on the beach, their party was disturbed by the arrival of three or four large dump trucks which replenished sand on the beach. Paradise, it seemed, had a commercial and artificial cover.

When they returned to the ship, it was humming with activity. Two mobile quarantine facilities (MQF) arrived for loading and storage. Creation of the MQFs was driven by a requirement from the Interagency Committee on Back Contamination (ICBC) to provide for containment of "moon germs" which might be carried back from the moon by the Apollo 11 astronauts.

The possibility that the CM itself might carry such pathogens back to earth was also considered, though some considered the possibility to be extremely unlikely. Several years prior to the flight of Apollo 11, NASA's Elbert King observed that the surface of the moon ought to be considered sterile, by earth standards, anyway. It was devoid of oxygen and water, which were necessary to support life. Ultraviolet radiation from the sun bombarded the lunar environment without mercy, without the benefit of filtration performed by an atmosphere like that on earth, and its surface temperature fluctuated from such extreme cold to hot that nothing commonly defined as life, on earth at least, could possibly survive the moon's environment. "If you really wanted to try to design a sterile surface," King commented, "this was it." But NASA bowed to pressure from the scientific community and agreed to develop both procedures and equipment to minimize the possibility that a moon-derived pathogen on board the Apollo 11 command module or carried by the astronauts themselves might infect the earth.

The ICBC agreed that the extreme heat of approximately 5,000° Fahrenheit generated during the fireball of re-entry to the earth's atmosphere would likely kill any such pathogens clinging to the spacecraft's exterior surface.

Apollo 11's astronauts followed a three-stage housekeeping regimen to cleanse the interior surfaces of the CM both prior to and during their return flight to earth, and prior to re-entry. That regimen was commonly referred to as "Brush, Dust, and Kick." After their walk on the moon, astronauts Neil Armstrong and Buzz Aldrin discarded some dust-covered items, like gloves and tools and boots, either on the moon or in the lunar lander itself. Then Armstrong and Aldrin vacuumed the interior of the lunar module, to cleanse it of any clinging moon germs. The Apollo 11 mission actually added one lunar orbit to their schedule to provide the additional time required to complete this action within the lunar module. Armstrong and Aldrin then docked their lunar module with the CM, flown at that time by fellow Apollo 11 astronaut, Mike Collins. An atmospheric pressure differential ensured that air flowed from the CM into the lunar module as Armstrong and Aldrin transferred to join Collins for the ride back to earth in the command module, *Columbia;* that air flow ensured that any remaining dust within the lunar module remained within that spacecraft and did not transfer to the CM. The entire process seemed a bit convoluted, but every procedure was driven by the ICBC's requirement to minimize the possibility that moon pathogens would somehow transfer back to earth. Finally, during their flight back to earth in the CM, Armstrong, Aldrin, and Collins wiped the interior surfaces of their spacecraft in a final effort to cleanse the spacecraft of pathogens. (Having taken all those precautions, however, sources later reported that when the CM was finally opened on board *Hornet*, its interior was covered with a fine layer of dust.)

In keeping with the ICBC's requirement, the astronauts would wear biological isolation garments, or BIG suits, upon egress from the module and would continue to wear them until they arrived on board *Hornet* and entered a specially constructed facility to begin a twenty-one-day period of quarantine while they were checked for indications that pathogens had returned with them from the moon. Two such facilities were loaded on *Hornet* in Pearl Harbor on Sunday, 6 July 1969.

NASA considered a contract to design and manufacture MQFs from scratch, but abandoned that idea after realizing that the cost to produce them would be prohibitive, easily more than a million dollars. The cost of the entire NASA manned spacecraft program, through the launch of Apollo 11, was approximately 24 billion dollars, so an additional million dollars to satisfy the ICBC requirements to prevent back contamination might have seemed reasonable to some. But NASA sought a cheaper, though equally effective route. So they contacted the AirStream trailer company—Airstream, Inc.—of Jackson Center, Ohio.

Airstream had been building quality travel trailers for customers since the 1930s. The bullet-shaped silver aluminum travel trailers had been towed behind family cars throughout the American highway system for decades, and were familiar to everyone. It so happened that their shape and size approximated NASA's requirement for a facility to house five people for a brief stay on board one of America's aircraft carriers after an Apollo splashdown at sea. If all went according to plan, such a facility would be required for only two days, the time required for a carrier to transit from a Pacific splashdown site to Hawaii, where the trailer could be transferred to a USAF C-141 cargo aircraft for transport back to the Manned Spacecraft Center in Houston, Texas. Upon arrival there, astronauts could be transferred from a trailer into a much larger facility inside one of NASA's labs. Quarantine inside a trailer, then, would not be forever, but most likely for a few days, at most.

For planning purposes, astronauts would have to be prepared to remain inside a trailer, on board the carrier, for the entire twenty-one-day quarantine period mandated by the ICBC. Such an eventuality might occur if the quarantine protocol was violated and the ICBC determined that the entire ship was possibly contaminated. In that event, the ship would remain at sea throughout the twenty-one-day quarantine period, and the MQF, including the astronauts, would remain on board.

Airstream, Inc., agreed to modify one model in their line of travel trailers to meet NASA and ICBC requirements, and they did so for a nominal cost. Four such trailers were then built by Airstream for NASA.

On Sunday, 6 July 1969, two of those specially modified trailers, designated MQFs, were loaded onto USS *Hornet*. The handling of the trailers proved both cumbersome and damaging. The rollers on the dolly used to move the trailers were too small, and they cut into the nonskid material that adhered to the deck, and into the wooden approach deck to the #3 elevator. But they made it into hangar bay #2, where they remained throughout the recovery operation.

NASA included an extra trailer for use by Navy frogmen or other personnel involved in the recovery and retrieval operation who might be inadvertently exposed to moon pathogens or contaminants during the operation. The second trailer remained empty and was not needed or used.

Commander Stuntz ensured that a wide white line was painted on the deck around the MQFs and the position nearby where the Apollo command module was expected to rest after its retrieval from the sea. He then posted two USMC sentries from the ship's MARDET to guard the perimeter and prevent sailors and civilian curiosity seekers from approaching those areas. Marines from the MARDET manned those posts twenty-four hours per day for almost three weeks, until the ship returned to Pearl Harbor from the recovery operation on 26 July 1969. Violation of the MQF boundary line could have resulted in a five-thousand-dollar fine or a period of imprisonment of one year. USMC sentries were armed, and had orders to lock and load in the event that anyone seriously challenged them. Naturally, someone would eventually do so.

6

Training at Sea

MONDAY, 7 JULY 1969

MEMBERS of the Hornet Module Retrieval Team had received their first formal training session, and then spent several days in port conducting walk-through exercises to familiarize themselves with Apollo retrieval procedures. It was time to try their hand at an actual retrieval. The launch of Apollo 11 was scheduled to occur in just nine days, so members of the retrieval team had their work cut out for them. Fortunately no other mission served as a distraction, and their commanding officer (CO) assured them that, with sufficient practice, they would develop the level of expertise both required and expected of them.

At 0719 on Monday, 7 July 1969, *Hornet* departed Pearl Harbor for three days of at-sea recovery training which began in an area approximately fifteen nautical miles south of Hawaii. Shortly after lunch that day, they began their first Apollo mock-up retrieval exercise, employing as a dummy target boilerplate #1218.

The USS *Hornet* Apollo 11 Cruise Report, an after-action report which summarized the three days of exercise from 7 to 9 July 1969, documented that nine practice retrievals were made during that three-day period, including six retrievals accomplished during daylight hours, and three at night. The officer in charge (OIC) of UDT Detachment

Apollo ensured that each of his swim teams was exercised during that period, and each swim team completed one day and one nighttime practice recovery. Weather and sea conditions varied throughout that period, with generally mild sea states, but with winds as high as twenty-four knots and six-foot waves, with light rain. Those conditions tended to complicate retrieval operations, but they also provided an opportunity to challenge the men. It was not easy. But the best lessons in life are often the toughest to learn. The sea accommodated those who looked for a good test.

Lt. (jg) Peter D. Beaulieu noted that at 0930 on the first day of operations at sea, the NASA team and the deck crew which would operate the B&A crane and the teams which manned lines to steady the boilerplate as she was hoisted aboard *Hornet* had their first meeting. A few hours later, at 1300, they conducted their first practice retrieval. That was just the beginning of an exhaustive two and a half weeks of practice sessions that would be conducted by the Hornet Module Retrieval Team, HS-4, and UDT Detachment Apollo before the actual recovery and retrieval of the Apollo 11 astronauts and their command module (CM) on 24 July 1969. Beaulieu stated that Captain Seiberlich "was interested in not dropping the ball in front of a record 500 million TV viewers, so the team practiced, and practiced, and practiced."

That first day, the team completed four boilerplate retrievals utilizing the B&A crane, including three during daylight hours and one at night. Mock astronaut recoveries by HS-4 and the UDT swimmers were also conducted. Debriefings were held after each exercise to identify both successes and failures, with a view toward boosting morale and correcting deficiencies. It was the classic carrot-and-stick method of training. The teams worked practically non-stop for twelve hours, with just one hour off for dinner between exercises.

To begin an exercise, crewmen rolled the boilerplate from its storage location in hangar bay #2. It was mounted on a specially constructed wheeled dolly for easy transport. Once positioned on elevator #3, it was raised to the deck and hooked to the B&A crane to be lowered into the water. The ship then slowly backed away from the floating boilerplate and removed itself by a distance of several miles, usually over the horizon and out of sight. The boilerplate remained alone on the sea, bobbing and traveling gently along the surface with the current.

The very first session lasted almost three hours, from 1239 to 1524. Lt. Shirley H. "Lee" Elliott served within the Combat Information Center (CIC) on *Hornet*, and he observed by manning radar scopes within the interior of the ship some of the difficulties and dangers which men occasionally encountered while performing recovery operations at sea. Once the boilerplate was positioned in the sea, its homing beacon activated and HS-4 helicopters employed their SARAH equipment to home in on the boilerplate's location. That's when the fun began.

Lieutenant Elliott observed that, when HS-4 helicopters acquired a bearing on the boilerplate's beacon, they tended to converge on that one point in the ocean, like steel balls to a magnet, at very high speeds. And that was dangerous. He recalled that they appeared on the radar scope like a bunch of young kids playing soccer for the very first time in their lives—instead of maintaining individual positions, everyone wanted to vector in toward the ball at once, because that was where the action was. And, back in the CIC, the action became confused and the plot became disoriented as all of the helicopters converged on that one spot. It was impossible to discern one helicopter from another, despite the application of new NASA gizmos to the IFF system. And since the CIC could not make sense of what was going on, their reports to the bridge were less than satisfactory. The first session appeared to be both dangerous and confusing. Navy radar operators were simply not used to the apparent undisciplined commotion, although HS-4 pilots had performed this type of exercise many times, as had the UDT swimmers. Their completion of mock recovery operations tended to go smoothly, from their perspective. It was the Hornet Module Retrieval Team which required the most work.

CWO1 Donald W. Harmer served as the ship's boatswain and was a key member of the Hornet Module Retrieval Team. The younger line officers on deck, including Lieutenant (jg) Beaulieu, considered Harmer to be particularly knowledgeable, experienced, and really good at solving seemingly difficult problems. He could be counted upon to get things done. The fact was, Harmer had simply "been there," in the Navy and working on ships, for a long time.

Harmer recalled that after the helicopters and swimmers completed most of their work, installing a flotation collar to the boilerplate and so on, the ship began its slow approach toward the boilerplate for retrieval

attempts. The approach had to be made from upwind, to avoid the possibility of contamination from "moon germs," and only after swimmers
decontaminated the mock CM and the astronauts—played by back-up
UDT swimmers—did *Hornet* then maneuver to approach the boilerplate
from the downwind side to begin retrieval operations.

Their initial failings were many and varied. The shot line fell short
of the boilerplate, causing the team to miss an opportunity to snag the
mock CM for a hoisting operation and forcing the ship to turn around;
a flexible NASA-supplied strap affixed to the end of the hoisting line for
retrieval operations came loose; a 1,500-pound test nylon line attached to
the sea anchor separated during in-haul, stretched and pulled too tightly
as the boilerplate was hoisted upward, and the whaleboat lacked a grap-
pling hook with which to retrieve it so the sea anchor was lost to the
sea—a significant event, since UDT teams brought only a limited number of sea anchors along; the whaleboat operated at such a great distance
from the swimmers that crewmen were out of position to assist with
shark watch duties; the whaleboat failed to maintain its position during
retrieval operations—it was supposed to position itself approximately 150
feet upwind of the module as the module passed down the starboard
side of the ship, in a position to retrieve the shot line if it overshot the
module; on one pass, during the ship's approach to the boilerplate for a
retrieval attempt, the ship passed outside the 400-foot range of the shot
line and had to again turn around, wasting time; a leak in one hydrau-
lic line on the B&A crane caused it to wind slowly; on one attempt, a
UDT swimmer signaled the okay to commence in-haul before detach-
ing the sea anchor, a situation which made the effort to haul the boilerplate close enough to the ship for a hoisting operation very difficult,
if not nearly impossible; the flotation collar bag, which was tethered to
the flotation collar itself, caught in the catchall nets beneath the elevator; during one attempt, the elevator was not in the full up position,
and when a large wave came along it lifted the boilerplate so high in the
water relative to the ship that the top of the boilerplate nearly struck the
bottom of the elevator—a strike against the elevator could have either
damaged an elevator track and put the elevator out of commission, or
the boilerplate could have been so damaged as to have been unusable for
additional practice sessions, a bad situation given the fact that only one

boilerplate was available for practice; a UDT life raft was lost to the sea after its tether line parted; confusion reigned for a time when personnel on the bridge employed a JV circuit (an internal telephone hard line between the bridge and elevator habitually used by Navy officers during maneuvering and docking procedures), rather than the NASA-preferred walkie-talkie communications link between the bridge, the elevator, and the whaleboat; the VHF walkie-talkie bandwidth was too wide, thereby allowing interference from the SPS-43 radar station to bleed into walkie-talkie communications from the bridge; and emanations from the Mutual Broadcasting van affected walkie-talkies used by teams on elevator #3 and the whaleboat.

CWO1 Harmer recalled that NASA supplied flexible nylon straps of varying lengths which affixed to the end of the hoist line. The flexible lines provided "give," or absorbed shock on those occasions when the 9,000-pound boilerplate would suddenly fall into a wave trough and pull against the hoisting line. Without the flexible straps affixed to the hoisting line to absorb that energy, the hoisting line itself would snap as the module fell away into the trough. Harmer learned that the rougher the sea state, the longer the strap required to offset tension on the line. NASA was additionally concerned that the suction effect of pulling the flat bottom of a 9,000- or 12,000-pound module directly up from the surface of the sea could—where the surface tension of the water might effectively "hold" the rounded bottom to the sea surface temporarily as the hoist line pulled upward which could also produce sufficient tension to snap that line—potentially lead to the loss of the craft altogether. The nylon straps at the end of the hoist line were designed to absorb that energy by stretching, and thereby preventing the line from parting.

BM3 Mickey N. Lowe was a member of the ship's 3rd Division (deck) selected to lead the team that removed the flotation collar from the boilerplate, and later the actual CM, after hoisting operations brought them out of the water and onto the ship. Lowe was selected by a senior chief in his division because he was considered by that supervisor to be his "go-to" guy to get things done.

Lowe's job on this detail was to cut the flotation collar loose from the spacecraft, and then assist his team as they lowered the craft onto a wheeled dolly for later transport back down to the hangar bay. Initially, the dolly was pulled by hand to its spot on the deck, and its wheels were chocked to prevent it from moving. But Lowe realized that *Hornet*'s pitching and rolling deck during heavy seas exceeded the chocks' ability to stay the dolly, which began to roll. Men and chocks alone could not stay the dolly, and given even heavier seas, the dolly could roll right off the deck and into the drink, to be forever lost. Lowe corrected the deficiency by the simple expedient of leaving the dolly hooked to a tractor, a "mule," which was ordinarily used by the Navy to move aircraft around the deck. It was a simple and effective solution to a very real problem that could have had serious consequences, and it was implemented by a junior enlisted man whose common sense and initiative served the best interests of the Navy and NASA. To Lowe, it was not a big deal. He was just doing his job.

The deck officers in charge of this operation, at the ground or sea level, were actually very young men, just a year or two out of college, with relatively little life experience under their belts, and with minimal experience in conducting operations at sea. They depended a great deal on the experience of senior enlisted men and warrant officers on the team who were seasoned veterans of the Navy. Lieutenant (jg) Beaulieu laughingly recalled that Lt. (jg) Paul Wineman served as officer in charge of the whaleboat. He recalls Wineman pleading, "Hey, you guys, when you drop us off with the boilerplate and then sail off over the horizon to begin your approach, the ocean looks a lot different from a whaleboat bouncing around. Don't get mixed up and forget that you left us out there. Okay, guys? Okay?"

Crewmen manning the in-haul line learned that danger often lurked at the end of that rope. An in-haul line was used to pull the boilerplate close in to the ship, so that a hoisting line could be attached to the boilerplate for a vertical hoisting operation. That seemed a simple enough concept, but they were working on the sea, where things can get complicated. The in-haul line was strung through a block attached to the deck. A team of men manned one end of that line, on deck, while the other end connected to the sea anchor ring on the module so that the team could pull the module inward toward the ship. But the team soon learned that the combination of a 9,000-pound craft and a heavy sea could produce dangerous consequences for the team. When a heavy wave or sea sloshed against the boilerplate, it moved the craft so suddenly that the line tended to pull away from the in-haul crew, pulling them as a body toward that block on the deck. If the first man in the line stood too close to the pulley, his hand would be pulled into the pulley and crushed.

Lieutenant (jg) Beaulieu observed that the small crew of men used to position the boilerplate dolly tended to stand around with nothing better to do until the boilerplate was once again ready to be placed on that dolly. So rather than have them stand around, he used them to augment the crew on the in-haul line, to add weight and strength to that line. He then made a note to himself: Everyone on the line was to wear life jackets, helmets, and gloves. Beaulieu recalled, "There were a million little details to attend to for an unfamiliar process. It was a learning process, and we learned by doing. There was a lot of downtime during practice sessions for one group or another; we took advantage of the downtime to brief and debrief the men, and to correct problems between practice sessions rather than wait for the end of the day" (author interview of Beaulieu, 5 September 2007).

Training lines were attached to the module as it was hoisted upward by the B&A crane, to keep it from swinging wildly to and fro as it was raised from the surface of the sea, and the men handling those lines on either side of the module displayed during those initial practice sessions a deep understanding of the word "teamwork" as they learned to handle the 9,000-pound swinging beast in heavy seas, while the carrier rolled and pitched in the dark. A NASA technician might brief them and tell them what to watch for and what to avoid, but until the crew developed some hands-on experience in handling a swinging 4.5-ton pendulum on a string, they would not really understand the problems or develop an ability to cope with problems as a team. In rolling seas, and with rain and wind spraying into their faces, they had to set that module just right on its dolly on deck, and they had little time for corrections or room for error.

A second practice session was held in mid-afternoon that day, from 1522 to 1723, and the third was held from 1847 to 2031. During the final practice session that day, which ran during hours of darkness until half-past midnight, from 2133 to 0033, the 24-inch carbon arc searchlight was used for the first time, having been commandeered by Electrician's Mates

Third Class Gasho and Laurence a couple of weeks earlier while *Hornet* was still moored in its home port of Long Beach, California.

TUESDAY, 8 JULY 1969

The next morning, *Hornet* steamed to a position approximately eighty nautical miles SSW of Hawaii, where she continued to conduct Apollo recovery and retrieval practices. Despite the many missteps and mistakes made during the previous day's efforts, the Hornet Module Retrieval Team already showed signs of steady improvement. Under the guidance of NASA experts, and given the benefit of learning while correcting mistakes and improving performance as they performed their jobs, the team's documented retrieval times, computed from the time that the boilerplate was within range of the shot line until the boilerplate was secured on its dolly on deck, ranged between nine and thirty-two minutes. The slowest time was attributed in large measure to poor weather conditions.

Multiple training exercises took place that day. In the morning, a flag officer arrived to gauge their progress. Rear Adm. Donald C. Davis, CTF-130, flew in by COD from Hawaii and landed on *Hornet*'s flight deck to observe the exercises. Two boilerplate recoveries and retrievals were conducted that day, utilizing the B&A crane and the NS-50 Tilley crane. The latter crane was mobile and positioned on the port side of the flight deck. It would serve as a backup to the preferred B&A crane during the Apollo 11 retrieval operation on 24 July 1969, in the event the B&A crane somehow became disabled, so exercising the men with the Tilley was necessary. A daylight exercise was held during a four-hour period from 0837 to 1235, and a brief nighttime exercise was held from 1904 to 2024.

Lieutenant (jg) Beaulieu was in charge of a deck team handling lines for the retrieval operation with the B&A crane. He recalled that the ship made a good approach to the floating boilerplate that day, passing within just fifty feet as the boilerplate slid down the starboard side. The distance was so short that a decision was made to refrain from firing a shot line from the ship to the UDT swimmers who accompanied the spacecraft. Instead, a monkey fist was simply tossed by hand to pass the in-haul line to the swimmers. It was a good decision, under the circumstances, but a key swimmer was momentarily hurt when the heavy fist hit him in the shoulder.

Monkey fists were used by sailors to pass a rope or line from one ship to another, or from a ship to shore, over a relatively short distance. Heavier gauged lines can be attached to a length of monkey fisted line; the recipient of the monkey fist can then use the small monkey fisted line to haul across that short distance the larger, heavier gauged line.

To enable a sailor to toss such a line for a distance of, say, fifty feet, a weight is attached to the end of the line, which is then heaved or slung with a coil of line—the attached line simply follows the weight to the recipient at the other end. The weight, commonly a small ball of lead, was affixed to the line by weaving or braiding the line around it, in the general shape of a small fist—a monkey fist.

Lt. (jg) Wesley T. Chesser was the leader of the three-man UDT swimmer team which worked with the boilerplate during the early recovery phase of the morning exercise. He had no direct role to play when the ship came alongside to retrieve the boilerplate, though, so he remained in a UDT raft which was tethered to the flotation collar around the boilerplate as it rode the current down the starboard side of *Hornet* for the retrieval exercise that morning.

Ordinarily, a shot line would have been fired from the ship to the boilerplate, then another designated swimmer on Chesser's team would retrieve the in-haul line. This line would then be attached to the boilerplate's sea anchor ring to allow the Hornet Module Retrieval Team to haul the boilerplate closer to the ship, directly beneath the B&A crane's dangling cable for hoisting. It was a routine which Chesser and his team had practiced many, many times as members of UDT Detachment Apollo. But the routine was broken that day. The decision to employ a monkey fist instead of a shot line was not clearly communicated to the swimmers.

Chesser was sitting in his raft when the monkey fist suddenly and unexpectedly hit him squarely in the shoulder. The impact shocked him at first and incapacitated him for a period of time. He was out of action until the pain subsided. The retrieval operation continued, and at the end of it, Chesser, who had recovered, joined his UDT teammates as they climbed up a vertical 50-foot length of cargo net, which had been draped over the side of the ship for the use of swimmers to board the ship.

The crew lost another sea anchor during the retrieval operation as its line separated during the hoist. The parachute-shaped sea anchor remained filled and weighted with seawater as it was dragged from the ocean upward during hoisting operations, and the strain of the load on the 50-foot tether was simply too great. It naturally broke. By the end of the second day of at-sea exercises, the crew had lost a total of four sea anchors in that manner. At that rate of loss, none would remain for the Apollo 11 recovery, and replacements would have to be priority requisitioned and flown halfway across the Pacific Ocean before splashdown.

Lieutenant (jg) Beaulieu noted that a simple solution was developed on the spot, in the form of a new procedure to be executed by UDT swimmers who accompanied the boilerplate throughout both recovery and retrieval operations. After the in-haul line was attached to the boilerplate during the hoisting operation, a UDT frogman would pull the sea anchor out of the water and simply dump it into an empty parachute mortar tube on the top of the boilerplate. With the sea anchor safely tucked away on top of the boilerplate, it would no longer break its tether during the hoisting operation. It was a clever solution to the vexing problem. It was better to make mistakes during practice sessions, Beaulieu thought, than to suffer them during the recovery of Apollo 11.

Lieutenant (jg) Beaulieu also noted that curiosity seekers were making a nuisance of themselves during the retrieval exercise. Ship's photographers and idle crewmembers who were not directly involved in the operation seemed to be running amok on elevator #3 as spectators, causing distractions and potentially endangering themselves, as well. The area was cleared, and orders issued to ensure that, in the future, onlookers maintained a safe distance from the operation itself.

Members of the Hornet Module Retrieval Team were engaged in serious business, and they were focused on their duties. But for many among *Hornet*'s crew, the entire assignment to support an Apollo recovery was exciting and new, and they simply could not refrain from gawking. Most of them were just kids.

Chief Warrant Officer 1 Harmer, the ship's boatswain, proved his worth during the exercise with the NS-50 Tilley crane. The Tilley was mounted on the flight deck, which was positioned approximately sixty feet above the waterline—quite a distance to hoist a nine-thousand-pound object from the water. As the Tilley began to hoist the boilerplate from the water, it began to twist uncontrollably. The problem lay with the housing for the Tilley hook, which spun as the crane hoisted the boilerplate. Men handling the training lines, who were charged with responsibility to steady the craft as it rose from the waterline, could not control the 9,000-pound monster as it twisted beneath the spinning housing. It was a dangerous situation. Harmer did not know why the housing for the Tilley hook spun as it did when it reached down to the waterline, but he figured a way to jury-rig the housing so that it would not spin. He welded one-half of a chain link to the housing and attached two separate lines to it. New teams of men were then positioned on elevator #2 to man those lines and to keep the hook and its housing steady as the crane hoisted the boilerplate from the water.

Again, a simple, clever, and effective solution to a vexing problem was developed. But only someone with years of hands-on experience, like Chief Warrant Officer 1 Harmer, could have thought of it and resolved the problem so quickly. Young Lieutenant (jg) Beaulieu's estimation of Harmer's value to the project went through the roof.

The daylight exercise that day went well and was completed shortly after noon. The next exercise, a nighttime exercise, was scheduled to commence some seven hours later, so, following their usual debriefing session, members of the Hornet Retrieval Team took a break.

The gap between the two practice sessions that day was filled with a STAR pickup from the flight deck. ABH1 Gary P. Righter had been in the Navy for many years by that point. He stated that he had never seen anything like it, though it was fun to watch. It looked like something straight out of a John Wayne movie entitled *The Green Berets*, which he had seen in a movie theater just a year earlier, 1968. Righter did not know exactly what was going on.

But Cdr. Irwin "George" Patch Jr., USS *Hornet*'s air officer, explained that it was just a practice pickup by a U.S. Air Force C-130 aircraft. The aircraft overflew *Hornet*'s flight deck and used a specially designed rig to scoop up materials from the flight deck as it flew by. It did not bother to land on *Hornet* and in fact was incapable of doing so. It simply flew by and picked up its cargo from the flight deck without stopping. The aircraft's mission in support of the Apollo 11 recovery was to retrieve press pool materials after the Apollo 11 splashdown—still photographs and videotapes, mostly—which would be transported expeditiously from *Hornet* to Hawaii for quick processing and further dissemination to the world after the recovery.

The system was originally called the Fulton Air Recovery System, named after its inventor, Robert Fulton—grandson of the man, by the same name, who invented the steam-powered ship. But the U.S. Air Force called it the Surface To Air Recovery, or STAR, system. Many called it the skyhook. A specially configured U.S. Air Force "Combat Talon" C-130—designated MC-130E—was used for the mission. Mounted on its nose were two "whiskers," shaped in the form of a V; the whiskers were about 12 feet long. At the center of the V was a latching mechanism that was designed to grab and retain a nylon rope as the aircraft flew by at a speed of 130 to 150 mph. Down below, on the flight deck of USS *Hornet*, a team stuffed the press pool's materials into a large duffel bag weighted to at least 100 pounds. Attached to that bag was a long nylon rope, 450 feet in length, and at the other end of the rope was a white balloon shaped like a blimp. As the aircraft made its approach toward *Hornet*, the team inflated the balloon with helium and released it to float hundreds of feet above *Hornet*. The pilot of the MC-130E then vectored his aircraft on an intercept course with the balloon and snagged it as he flew by. The latch at the center of the V captured the nylon rope and simultaneously cut the balloon free to float off into the atmosphere. Crewmen on board the aircraft then winched the duffel bag into the cargo bay of the aircraft.

In that manner, news photographs and videotape of the Apollo 11 recovery and the follow-on welcoming ceremony would be expeditiously transported from *Hornet* (which would then be operating approximately one thousand miles south and west of Hawaii) to laboratories on Hawaii for processing and further dissemination to an eager public.

It was a remarkable system. Its inventor, Robert Fulton, also invented the system employed by UDT swimmers to board fast boats at sea as they treaded water on the surface. And he inspired the system employed by U.S. Air Force para-rescue swimmers to deploy flotation collars and related recovery materials to NASA spacecraft after splashdown. That system, called the Aircraft Deployed Drift Reduction System (ADDRS), was on board U.S. Air Force HC-130 aircraft flown by the 41st Aerospace Rescue and Recovery Wing. Two such aircraft were positioned hundreds of miles to the north and south of Apollo 11's primary landing area, in case the module fell short of, or overshot the designated splashdown point.

Robert Fulton's contribution to the successful recoveries of NASA manned spacecraft may not be generally or widely known. But he is one of the unsung heroes of the era.

WEDNESDAY, 9 JULY 1969

One week remained before the scheduled launch of Apollo 11. USS *Hornet* continued to operate in fairly close proximity to Hawaii, just seventy-five nautical miles SSE from the island chain.

The ship conducted just one exercise that day, but it was the team's very first full-scale simulated exercise, or SIMEX 1, so it represented an important milestone in their training. The exercise was timed to closely coincide with the anticipated schedule for the return of the Apollo 11 command module, *Columbia*, to earth. It ran for approximately three and a half hours, from 0445 to 0821.

By 0445, all aircraft designated to support the SIMEX were aloft. It was still dark. They included two E-1B aircraft, which performed air boss and relay functions; four HS-4 Sikorsky SH-3D helicopters, including the recovery aircraft and three UDT swim teams; and two U.S. Air Force HC-130s, which operated at distances of approximately 165 nautical miles north and south of the simulated re-entry track. U.S. Air Force para-rescue swimmers were on board their aircraft.

The wind blew at a moderate ten to fifteen knots in the exercise area, and the sea state offered acceptable three- to five-foot waves on top of swells. The weather was absolutely perfect for an Apollo recovery, though possibly less challenging than NASA and Navy personnel might have preferred for an exercise. Lieutenant (jg) Beaulieu took notes and continued to list and correct deficiencies as the exercise progressed, but there were few missteps along the way.

UDT swimmers failed to secure a package of material to one of their rafts during the boilerplate retrieval operation, and the package was lost at sea. HS-4 helicopters positioned in standby status on the flight deck of *Hornet* were staged in the same general area to be used by presidential helicopters during the Apollo 11 recovery operation; an E-1B flew above the flight deck to eyeball the problem, with a view toward developing a solution to deconflict those spacing requirements. And walkie-talkies used by persons manning the whaleboat, the bridge, and elevator #3—all of whom necessarily coordinated their actions during the retrieval operation—used a frequency which was so close to that used by an HS-4 helicopter pilot to talk to someone else, that the communications often overrode or otherwise interfered with one another.

Some of the little kinks had yet to be worked out. But all in all, the SIMEX went smoothly. During the debriefing which followed the exercise, Dr. Stullken, the chief of NASA's onboard team, proclaimed, "This is the best team yet. Just keep it up. You are far ahead of all other retrieval teams at this stage of the game." And Captain Tollefson of CTF-130 described the exercise as "basically the best simulation that I've seen." Lt. (jg) Richard E. Powers III was one of the young, twenty-three-year-old novice Navy officers assigned to *Hornet* during the Apollo 11 recovery. His duties included interaction with members of the press as well as the NASA team on board ship. He recalled the following comment which issued from a NASA employee after the debriefing, made as an aside: "Here we are on the biggest of them all, and we drew a Captain whose performance in preparing the ship is the best in the history of the program."

Captain Seiberlich's efforts were not going unnoticed or unappreciated. Nor were those of his crew. But all was not well. Though the crew of *Hornet* and other Navy personnel assigned to support the Apollo 11 recovery effort certainly seemed to have risen to the challenge, the ship itself, then twenty-six years old and nearing the end of its expected life cycle, coughed and sputtered just a bit. MM1 Ernest E. Lee recalled that one of the ship's four engines shut down during the exercise as a bearing on a propeller shaft burned out.

Hornet's engines produced energy to drive the propeller shafts. When an engine stopped, the ship lost power and consequently lost the ability to move through the water at high speed. Fortunately, the ship was scheduled to return to Pearl Harbor that afternoon anyway, and Pearl Harbor was relatively close by. The decision to make repairs in Pearl Harbor rather than at sea was an easy one.

USS *Hornet* returned to Pearl Harbor at 1325 on Wednesday, 9 July 1969, and moored at Ford Island with its starboard side to Pier B, berths 25 and 26. It would remain in Pearl Harbor for just two and a half days, until its departure on Saturday, 12 July 1969, and would not return to Hawaii thereafter until after the recovery of Apollo 11 astronauts and the retrieval of their command module, *Columbia*.

7

Final Preparations

THEIR first night in port was not without its distractions. MM3 Robert W. Zengerle was a young machinist's mate assigned to the same division on *Hornet* as MM1 Ernest E. Lee. He worked with the engines. A few nights before *Hornet* left Pearl Harbor to conduct at-sea recovery and retrieval exercises, Zengerle and some of his buddies spent time in downtown Honolulu on liberty, as sailors often do while in port. Honolulu was attractive, and the city tended to cater to Navy men. There were plenty of bars available in which to pass the time. But it was not uncommon for an otherwise tranquil, if not idyllic, night on the town to turn into a scene of mayhem. And sure enough, fights between blacks and whites did break out in the bars downtown while Zengerle was there. He steered clear of trouble, however, and made it back to the ship unscathed. Upon his return to Pearl Harbor from several days at sea, then, Zengerle decided to avoid the downtown area altogether and, with luck, thereby avoid that kind of trouble before the ship departed in a couple of days to pick up the Apollo 11 astronauts. He opted instead to spend some time at the enlisted men's club (EM club), on base. That seemed a safe bet.

The club closed around two in the morning, and Zengerle milled about for a few more minutes with the crowd. Eventually, though, he began to make his way back to the ship on foot, alone. Zengerle walked two to three blocks away from the club and then spotted a small group of black sailors headed toward him, walking in the opposite direction. There were perhaps fifteen of them, and he thought they might have been assigned to the base. As they approached, comments were made by members of the group which were less than complimentary about the white sailor, and a fight broke out. Zengerle recalled that they were in a very narrow walkway between stacks of material stored on the pier, and though members of the group might have wanted to get at him, they simply could not do so, more than a few at a time. He got some licks in and took more in return, but the fighting was in such a confined area that neither Zengerle nor members of the group could get a good swing in. Zengerle quickly sought refuge under a nearby stack of pipes on the pier and squeezed as far beneath the stack as possible. The group spent some time kicking at him, but eventually they declared victory and went on their way. *Hornet*'s deck log recorded at 0315 that a fight involving Zengerle had broken out on Pier B, and that he had been taken to the dentist for treatment. Zengerle reported for duty on time that same day. A rumor spread throughout the ship that a *Hornet* crewmember had killed another sailor during a fight on the pier earlier that morning, but that the *Hornet* sailor avoided charges because he had fought in self-defense and, further, that *Hornet* was consequently ordered to depart port the following day. Zengerle laughs about the incident today, and dismisses the rumor as typical scuttlebutt on a Navy ship.

Hornet's XO routinely dealt with similar incidents, as the officer in charge of day-to-day operations of the ship. Racial tensions on board ship, particularly those between black and white sailors, mirrored those which were common to American society as a whole during the mid- to late 1960s. The discipline and good order of the ship's crew was never seriously threatened by the atmosphere, and the ability of *Hornet* to accomplish its mission was never in question, but it was an issue which required the XO's time and attention, and his occasional intervention.

A number of black sailors on the ship had formed an informal group which adopted the moniker "The Black Bullets." They threaded what resembled black bullets onto their dog tag chains and wore them around their necks, and occasionally caused problems during leisure time on board, especially around the basketball court in the hangar bay. Lamb developed a small network of sources among the crew who occasionally alerted him to Black Bullet activities, and he acted on the information to keep troublemakers off balance, and occasionally to humor them. If they held a secret meeting somewhere deep within the bowels of the ship, Lamb would make a point of casually walking into the area, much to their surprise and dismay, just to join their meetings. If Black Bullets were hogging the basketball court, to the exclusion of white sailors, Lamb would don his sweats and join them for a game—pointedly informing certain players that they appeared to be tired from their strenuous efforts on the court and that they ought to sit out and allow the XO and others to play; then he would substitute white players into the game, until everyone understood that the XO expected them to play nicely together. Lamb had a sense of humor and a very approachable manner, but the men understood and respected him. There were no major problems. Racial tension was occasionally in the air, but always in the background, and he would not allow it to interfere with the operation of the ship.

While Zengerle received treatment from the ship's dentist that morning, the ship's daily routine continued to march apace. The ship's surgeon, Lt. Donald Giulianti, USNR, detached from *Hornet* with orders to report to the Naval hospital at the USMC base in Quantico, Virginia. He would therefore miss his opportunity to participate in, or at least to observe, one of the most significant events of his generation. But his replacement, Lt. (msc) Stephen A. Habener, USNR, happily reported for duty that same day, just in time to shove off for the recovery, and he appreciates to this day how fortunate he was to do so.

At Pearl, routine loading and resupply operations were conducted, and Machinist's Mate First Class Lee's bearing was replaced. Once again, *Hornet* was prepared to drive full speed ahead, if necessary, and she was ready for action. Lieutenant (jg) Beaulieu noted that the crew practiced dry runs of the off-loading procedures for the MQFs and the command module. They did so, in preparation for the day that *Hornet* returned to Pearl Harbor after the recovery to off-load the quarantined astronauts onto the pier for transport back to Houston and to off-load the command module, *Columbia*, for processing and eventual transport to Houston, as well. A welcoming celebration for the astronauts was expected to take place on the pier immediately after the ship returned to Pearl Harbor with the Apollo 11 astronauts, and Captain Seiberlich wanted no glitches to spoil the party—so his men practiced every step of the off-loading procedures, just to ensure that an error-free celebration would be enjoyed by all.

Lt. Cdr. Joseph A. Fidd, *Hornet's* supply officer, noted that while he was engaged in the practical exercise of loading supplies onto the ship, Captain Seiberlich busied himself by hosting visits by various ladies' groups from Honolulu and the Navy base who wished to share in the Apollo 11 experience and who also wished to convey to the commanding officer of the Primary Recovery Ship for the recovery of Apollo 11 their heartfelt desire for fair winds and following seas during the recovery operation. Both men accomplished missions in support of USS *Hornet* operations that day.

FRIDAY, 11 JULY 1969

This was *Hornet's* last full day in port before the ship departed to recover Apollo 11. And it was a very busy day, indeed. A check-in/check-out desk was established on the quarterdeck to process approximately 35 NASA personnel who would support *Hornet* during the final leg of its journey, and approximately 120 reporters who would board *Hornet* to observe and report to a waiting world the details of the Navy's efforts to achieve the second half of President John F. Kennedy's stated national goal.

The first half of that goal was to land a man on the moon before the decade's end. That portion of the goal would be met in just nine days, on 20 July 1969, when civilian astronaut Neil A. Armstrong leaped from a ladder mounted beneath a crew hatch on Apollo 11's lunar lander, *Eagle*, placing his boots on the surface of the moon. The second half of the president's goal—constituting the very core of *Hornet's* mission—was to return that man safely to earth.

As they checked in, each civilian guest received a name tag for easy identification, a special VIP kit that included a commemorative Apollo 11 baseball cap, an Apollo 11 ballpoint pen, a collection of sundries, and "Captain's Cover" envelopes—commemorative envelopes which featured Apollo 11–related sketches drawn by a member of *Hornet's* crew, Lieutenant (jg) Beaulieu. And they received a "welcome aboard" brochure which provided information about life on board Navy ships. In order to support the effort to accommodate the large number of civilian guests now reporting on board, the Navy temporarily assigned to *Hornet* an additional twenty-five enlisted stewards who assisted with luggage and answered questions as the guests settled into their staterooms.

Cdr. Irwin "George" Patch Jr, *Hornet*'s air officer, was responsible for finding rooms for the guests. Patch and Cdr. John J. "JJ" McNally, *Hornet*'s operations officer, were classmates at the U.S. Naval Academy (class of '53), and they interacted at the academy with Midn. Johnny Stuntz, brother of weapons officer Cdr. Harley L. Stuntz. Theirs was a small and insular world. Patch was all Navy and he came from a military family. When the Cuban Missile Crisis broke out in October 1962, less than seven years earlier, the Department of Defense had available for planning purposes only some outdated maps of the island—maps which were developed for the Army by Patch's dad.

Fortunately, *Hornet* had left the bulk of its air arm back in California to free up room on the ship to accommodate guests, making officer staterooms available, though they were austere by civilian standards. Civilians were generally not accustomed to life on board a Navy warship. A two-week stay on board *Hornet* was not anything like a vacation on board a commercial cruise ship. Staterooms were sparsely furnished, small, and crowded; furnishings included a small bunk for sleeping, a locker, drawers, and if lucky a small desk and washbasin. The rooms were built to be functional, not comfortable, and all of the civilians on board for the recovery shared staterooms with several other men. The rooms were painted a Navy gray and had no curtains, decorations, or pictures on the bulkhead. Curtains would have been superfluous in any event, as the staterooms lacked portholes. They also lacked private bathrooms. There would be no room service.

But the ship did offer some amenities and services for their guests. U.S. Navy enlisted personnel under the direct supervision of SDCS Charles E. Raigans assisted guests and performed personal errands on request. Laundry service was provided twice per week, but not dry cleaning; turnaround time for laundry was forty-eight hours. The Navy even embarked from the commander of Naval Air Forces Pacific, based in Hawaii, a special complement of ten laundrymen to perform the extra work. The ship also served three meals per day—cafeteria style for breakfast and lunch—sit-down dinners were served in shifts, in the flight mess located one deck below the officers wardroom. A snack bar served hot and cold sandwiches, beverages, chili, and "specials," at cost, from 2000 until 0100 in the morning.

For entertainment, movies were shown in the dining room twice each night; they were rarely "first-run" movies, but rather old classics. The ship's radio station was on the air from 0600 to 2400 daily. TV3, a closed-circuit TV network on board *Hornet*, operated from 1600 to 2200 daily and presented full-length movies, documentaries, and interviews with key recovery personnel. The ship also produced a daily newspaper. And, following lunch and dinner each day, the Commander Naval Air Pacific (COMNAVAIRPAC) band, based in Hawaii but detailed to *Hornet* for the astronauts' welcoming ceremony on 24 July 1969, provided concerts. But Lt. Leroy Henry "Roy" Knaub, a qualified *Hornet* COD pilot, recalled that, at night the most common form of entertainment among the civilians on board *Hornet* was poker, played in an officer's stateroom, where players easily won or lost sums up to thirty dollars per night.

Civilian guests were permitted to purchase items from *Hornet*'s exchange store—except for uniform items—for cash only; checks were not accepted. *Hornet* additionally offered a tailor shop, a cobbler shop, a barber shop, a post office, a library, medical care and emergency dental care, and a chapel which offered separate Protestant, Jewish, and Catholic services. The welcome aboard packet specifically informed guests that sea sickness could be treated in the sick bay on the second deck amidships.

And civilians could place telephone calls back home. ABC, GE, and the Mutual Broadcasting System established a total 530 telephone patches to enable both civilian and military personnel on board *Hornet* to place long-distance telephone calls back to the United States. The USMC provided a special communications group to augment *Hornet*'s telephone exchange to handle all the calls. *Hornet*'s final cruise report recommended future assignments for the Marine Corps team, because USMC personnel were uniformly "cooperative, proficient, and motivated."

But there were also a host of inconveniences for civilian guests, which were actually the norm for sailors at sea. Guests would be required to take Navy showers, which were commonly called "Marine" showers. Given the need to conserve the limited supply of freshwater on board ship, guests were cautioned to follow a strict regimen when showering: turn on the water only long enough to wet the body; turn off the water; apply soap; turn the water on only long enough to rinse, and then shut it off. A typical Navy shower might last only a minute or two. ETR3 Dennis R. Moran recalled that many of the civilians on board *Hornet* seemed to ignore the rules regarding the conservation of freshwater and opted instead to take their usual "Hollywood showers." Such an infraction by a sailor would have resulted in an instant reprimand and the administration of rather harsh sanctions, though peer pressure alone usually sufficed to govern the issue; violations among the crew were practically nonexistent. As it was, Moran recalled that the ship often resorted to "water hours" during the Apollo 11 recovery effort to compensate for excessive consumption of freshwater by the civilians on board. During water hours, bathing was prohibited altogether until a sufficient reserve of freshwater could be produced by the ship.

Smoking was restricted. Guests were informed that fire aboard a Navy ship could spell disaster. Restrictions were numerous, widespread, and strictly enforced. An individual could not simply walk around a Navy ship with a cigarette hanging out of his mouth. If the word "the smoking lamp is out throughout the ship" was passed over the public address system, then all smoking must cease instantly.

Alcohol was strictly forbidden aboard naval vessels, and civilians were no exception to that rule, first implemented in 1914 by Josephus Daniels through General Order 99, which banned the substance from all ships in the U.S. Navy. The U.S. Navy has been "dry" ever since, and the most common form of beverage consumed on U.S. Navy vessels since the issuance of General Order 99 has been coffee. Sailors ironically dubbed the latter beverage a "cup o' Joe" in honor of the man who reduced their choices of drink while at sea to coffee or water. *Hornet* was so intent on clearly communicating their views about the consumption of alcohol on board ship that they included a feature in the welcome aboard packets: a sketch of an outstretched palm hovering above and covering entirely the lip of a whiskey glass and below, the caption "The ultimate sacrifice . . . self denial."

But that prohibition reportedly held about as much water for *Hornet*'s guests as the Marine showers. JO2 Chauncey "Chan" Cochran was an enlisted journalism specialist who was detailed on temporary duty orders from the office of the Commander in Chief of the Pacific Fleet (CINCPACFLT), based in Hawaii, to support the public affairs office on board *Hornet* during the effort to recover Apollo 11. Chan's job was to assist members of the press pool in the filing of their stories, to ensure that the ship transmitted them in a timely manner. He thereby developed a very close working relationship with certain members of the onboard press pool. Chan recalled that *National Geographic* photographer Bob "Twiggy" Madden brought on board approximately twenty foot lockers full of equipment, including forty to fifty Nikon cameras he set up throughout the ship to take still shots during the welcoming aboard ceremony for the astronauts. But one of those lockers was actually filled with bottles of Jack Daniels, individually wrapped in cotton socks to prevent breakage. The report was not rumor or "scuttlebutt." Cochran would later risk judicial punishment by imbibing in a few celebratory drinks himself.

Ultimately privacy would be sacrificed almost entirely. Civilians shared staterooms, bathrooms, dining facilities, and virtually everything else. And the constant inconvenience for everyone on board ship was incessant noise. Aircraft carriers are working ships and places of constant commotion, especially during daylight hours. The din of engines and vibrations and the background noise of work throughout the ship never ended. For all of this, civilians were charged $2.50 per day for their room, board, and professional upkeep.

Civilian reporters learned that the Navy provided not only some small comforts of home throughout their stay on board ship, but also a large measure of professional assistance. *Hornet*'s public affairs officer (PAO), a former Stanford University heavyweight wrestler, Lt. (jg) Milton "Tim" Wilson III, and his staff were supported during the trip by temporary assignees from throughout the Pacific theater to ensure that the media were well served in their efforts to report details concerning the recovery effort. Their media center was located on the centerline passageway on the O-2 level of the ship, in close proximity to Ready Room One, which was used by members of HS-4 and UDT Detachment Apollo. Press conferences were scheduled twice per day, at 0800 and 1600. PAO staffers could schedule and arrange interviews with members of the crew and dinners with Captain Seiberlich.

Ens. William C. "Bill" Whitman reported for duty on board *Hornet* in May, shortly after she returned from her final WestPac cruise. He was, in fact, the very last line officer to report for duty on *Hornet* before she was decommissioned about a year later in June 1970. Whitman was assigned to assist *Hornet*'s PAO by serving as a liaison officer for the press pool. His duties included assisting with the scheduling and arranging interviews with members of the crew, answering questions posed by reporters about shipboard operations, and collecting facts about those operations for the press pool.

Lt. (jg) James E. Ovard served both as the embarked communications liaison officer to coordinate berthing for the press corps and as an alternate point of contact for members of the press for anything else they might require. The Apollo 11 recovery would be his swan song in the Navy, his final duty—he mustered out of the Navy about two weeks after the ship returned to Long Beach following the recovery. Another young *Hornet* officer, Lt. (jg) James S. "Steve" Lauck, recalled that Ovard aspired to serve in broadcast communications some day, so the opportunity to work with reporters on board *Hornet*, and particularly to work with broadcast reporters, excited him.

The final loading of stores and equipment were completed that day, Friday, 11 July 1969, the day before *Hornet* was scheduled to depart Pearl Harbor to accomplish the mission to recover Apollo 11. She was ready. But Lieutenant (jg) Beaulieu, one of the deck officers charged with responsibility for accomplishing that mission, noted that the Hornet Module Retrieval Team suffered a significant loss that day when the 1st Division's leading petty officer, BM1 Roy L. Wallace, was suddenly sent home on emergency leave.

Wallace was one of those go-to guys who could seemingly do anything. He had received a formal education only through the seventh grade while growing up somewhere in Mississippi, but he was an intelligent, street-smart, very capable member of the crew who simply knew how to get things done and who exercised the personal initiative to do so. Wallace was a feisty guy who had once earned the title of heavyweight boxing champion for Navy forces in the Western Pacific and had served as the right-hand man for ship's boatswain's mate, CWO1 Donald Harmer, during practice sessions for the retrieval operation. The team would manage, but Wallace would be missed.

Somewhere along their path to success, Captain Seiberlich developed the idea to adopt a ship's motto to capture the very essence of *Hornet*'s mission to accomplish the second half of the national goal set by President Kennedy: to bring the Apollo astronauts safely back to earth. A motto could build team spirit among the crew, focus their efforts and unify them in a determined effort to accomplish a mission that would have immediate national and international impact, given its high visibility around the world, and would accrue a greater value in history as time went by.

A contest was held among the ship's crew. A bright young sailor whose identity has since been lost to memory won that contest with the motto "Hornet + 3." It summarized in the most simple terms what the men of USS *Hornet*, NASA, HS-4, and UDT Detachment Apollo were determined to accomplish in the Pacific Ocean on 24 July 1969: to sail to a splashdown point in the ocean as a dedicated and unified team, and then, simply to turn around and bring three additional men, the American astronauts of Apollo 11, safely back home.

8

UDT Detachment Apollo

W HEN the Navy accepted the mission from NASA to support recovery operations at sea, it understood that it had to plan for the worst-case scenario and then simply hope for the best. Competent swimmers were abundant in the Navy. There were lifeguards and swimming instructors, deep-sea divers, underwater construction specialists, explosive ordnance disposal experts, and so on. The water, after all, was the Navy's element. But the Navy turned without much debate to the frogmen of their underwater demolitions teams, UDT.

At the time, the Navy maintained three UDT teams on the West Coast of the United States, UDT-11, UDT-12, and UDT-13; all were based at the Naval Amphibious Base Coronado, California. Each team consisted of approximately 25 officers and 111 men, when fully staffed. Throughout the period of America's involvement in the Vietnam War, the teams rotated at six-month intervals in and out of Vietnam for combat tours. Thus a team could expect to spend a period of six months in Vietnam, followed by a year back at their base in Coronado, and then return for another six-month tour in Vietnam.

When teams returned from Vietnam to Coronado for their one-year tour of duty in the United States, they typically broke up into separate and smaller detachments or platoons to perform specialized assignments. Thus one detachment might work with the development of minisubmarines (or SEAL delivery vehicles [SDV]); another, the development of new tactical weapons; another, procedures for delivery of tactical nuclear weapons by UDT/SEAL forces; and another—UDT Detachment Apollo—support for Apollo recovery operations.

UDT-11 completed a tour in Vietnam at the end of 1968 and returned to Coronado to begin a yearlong tour of duty stateside. Twelve men were selected for the detachment which supported NASA's manned spacecraft program. Their detachment commander was a twenty-five-year-old officer named Clarence James "Clancy" Hatleberg.

Hatleberg, the son of a World War II Navy surgeon who served with the Marines at Iwo Jima, grew up during the 1950s in the small farming community of Chippewa Falls, Wisconsin. Thirteen years old when the Soviets launched *Sputnik,* he was immediately drawn by the allure of space travel. After high school he attended Dartmouth, where he joined the NROTC, spending his summers on destroyers and air craft carriers and attending basic flight school before graduating in 1965 and making a career in the Navy. He graduated in 1966 from UDT Training (UDTRA) class #39 and served for nine years with Navy underwater demolition teams; his last assignment with the teams was as commanding officer of UDT-11. He spent two tours of duty in combat as a UDT officer, serving in the Mekong Delta in Vietnam, and was wounded by shrapnel, pieces of which were never removed from his body. He was a veteran of the Apollo missions and a participant in the recovery of Apollo 6. In May 1969, he led the team that recovered Apollo 10 and later provided instruction to the UDT team responsible for the recovery of Apollo 12.

Hatleberg was a senior lieutenant when he assumed command of UDT Detachment Apollo in late 1968. The detachment consisted of twelve men: Clancy, three junior officers, and eight enlisted men. He divided his detachment into three-man swimmer teams and created billets for two additional swimmers to serve as alternates to substitute for sick, injured, or otherwise absent swimmers, as necessary. Very soon, one three-man team within UDT Detachment Apollo led by Lt. (jg) Wesley T. "Wes" Chesser emerged as more experienced and consistently excellent in executing the recovery procedures than others. Subordinate members of Chesser's team were QM3 Michael G. Mallory of Aldelwood Mannor, Washington (near Seattle), and SN John M. Wolfram of Fort Atkinson,

Wisconsin. The team of Chesser, Mallory, and Wolfram was designated as UDT Detachment Apollo's primary recovery team, the most experienced and capable team. In them, Hatleberg had tremendous confidence.

Mike Mallory and John Wolfram were more than dependable teammates. Like his commanding officer, Mallory also served in Vietnam. Much of that time had been spent deployed on board USS *Tunny* (SS-282), a highly decorated World War II–era diesel-powered submarine that was specially configured to support Navy Special Warfare operations. From *Tunny*, Mike and his unit performed beach reconnaissance and a record number of beach survey operations. Wolfram was also a seasoned member of the group, despite being only twenty when he became a member of the Apollo 11 recovery team. Wolfram served one combat tour in Vietnam prior to his assignment to UDT Detachment Apollo and followed that assignment with yet another combat tour in Vietnam. Indeed, because the U.S. Navy cared enough about the success of the mission to recover the astronauts of Apollo 11, Hatleberg, Chesser, Mallory, and Wolfram were just four of the twelve UDT frogmen sent to support that mission. Each man in the UDT detachment was exceptional.

Chesser's team, having been designated by the officer in charge of UDT Demolition Apollo as his primary recovery team, did not mean that that team would necessarily "get their feet wet" to accomplish the recovery. The governing principle established by NASA, and to which HS-4 and previous UDT teams had adhered throughout the manned spacecraft program, was that the first of the three UDT swim teams to reach the spacecraft after splashdown would be ordered to deploy. The others would hover within their helicopters in standby positions until called on by the deployed team for assistance. If no such call was issued, then the other two UDT swim teams would return to the primary recovery ship (PRS) without getting their feet wet at all, unless they had an opportunity to recover a drogue parachute, a main parachute, the Apex cover, or some other item that shed from the command module (CM) during its descent. The bottom line was that, preferences of the OIC of UDT Detachment Apollo did not carry much weight in the equation. The commanding officer of HS-4, piloting the Recovery helicopter, Cdr. Donald S. Jones, would make the call.

Chesser's team did deploy for the recovery of the Apollo mission in late May 1969—that of Apollo 10—just a week or so before USS *Hornet* was selected to serve as PRS for Apollo 11. Chesser's sea anchor man for the Apollo 10 recovery had been a swimmer of French-Canadian descent named Louis Boisvert. Boisvert dropped from the program after the recovery of Apollo 10, for personal reasons; Wolfram was assigned to the sea anchor spot on Chesser's team.

Boisvert was not the only UDT frogman to rotate out of UDT Detachment Apollo to accept another assignment or to move on for some other reason. To the men of UDT-11, and in particular to those of UDT Detachment Apollo, it did not matter. From their perspective, every Navy frogman was exceptional, and certainly qualified to support NASA's manned spacecraft program after completing specialized training in recovery operations. As for Boisvert, he continued to operate with the teams for twenty-eight years, first as a frogman and then as a SEAL.

For the recovery of Apollo 10, Chesser served not only as the leader of his specific team within the detachment, he also served as the acting OIC of the detachment. Clancy Hatleberg was given temporary orders at the last minute to report to Houston to participate in BIG training—Hatleberg and the astronauts donned biological isolation garments (BIG) to mimic procedures that would be followed after splashdown. In his absence, Chesser assumed a leadership role.

The recovery of Apollo 10 was successful, spectacularly so. The weather in the Pacific happened to be perfect—sunny, clear, and calm. And the CM landed right on target, less than two miles from the ship. At sea, two miles seems like right next door. The UDT team worked efficiently and almost everyone was happy with the outcome of the recovery mission.

But Dr. Donald E. Stullken was not a happy man. Dr. Stullken was the chief of NASA's Landing and Recovery Division; he also served as leader of the NASA support team on board USS *Princeton* (LPH-5) for the recovery of Apollo 10. NASA held him ultimately responsible for the success of the mission. So when Dr. Stullken learned that an unauthorized "sticker" of some kind had been glued to the 9-inch window of the side crew hatch on the Apollo 10 command module as it was being winched aboard ship, he was not amused. The sticker was a colorful flower appliqué commonly sold in bed and bath stores as a decorative non-skid material to prevent falls in the shower or bath. Application of the sticker to the Apollo 10 command module was just a prank. But it was not funny

Dr. Stullken grabbed a bullhorn and addressed the assembled multitude of Navy, NASA, and media civilians gathered in *Princeton*'s hangar bay to observe the final stages of the retrieval effort. He asked pointedly for the culprit to show himself. No one stepped forward, but Stullken's request ignited a wave of snickers and stifled laughs from within the crowd. Stullken then explained the basis for his ire.

The purpose for retrieving a CM from the sea after splashdown, he explained, was not for NASA to collect a memento from its Apollo mission. The retrieved CM was not simply some kind of NASA souvenir destined, perhaps, for a museum somewhere. The purpose, instead, was to make the CM available to NASA scientists and engineers, including contractors who built the machine, for examination, with a view toward determining everything that happened to the spacecraft during its flight. Lessons learned through such examinations would then be applied to spacecraft designated for use during future flights, to enhance the safety of astronauts who have to fly the machines. When someone "slapped a sticker" on the window of Apollo 10's side crew hatch, evidence of what actually happened to that window was destroyed. He was not happy. Accusatory fingers were not pointed in their direction, but insofar as UDT swimmers were in the water with the module just before it was hoisted onto the deck, the application of a small measure of deductive reasoning led some to conclude that a swimmer might have been culpable.

Frogmen were reputed to be great pranksters anyway. And they were consequently the source of endless headaches for Dr. Stullken, who referred to them collectively as a "pain in the ass" because of their high jinks during recovery operations. It was Stullken, after all, who, as the senior NASA official on board PRS during Apollo recovery and retrieval operations, had to answer to NASA superiors back in Houston for any failure to control the high-energy, rambunctious frogmen of UDT.

While Wesley Chesser was dealing with matters in the Pacific in late May 1969 to ensure a successful recovery by UDT Detachment Apollo of the CM for Apollo 10, his boss, Clancy Hatleberg, had BIG matters to attend to in Houston. Hatleberg had never met the Apollo 11 astronauts before. Although he was familiar with the BIG suits, which both he and the astronauts would wear during the recovery operation after splashdown to guard against the spread of moon pathogens, and although he had studied the procedures he was required to follow to disinfect both the astronauts and portions of their spacecraft before HS-4's Recovery helicopter transported the freshly cleansed astronauts to *Hornet*, Hatleberg and the astronauts had never practiced those procedures together as a team. Someone in Houston must have realized the oversight at the last minute and ordered Hatleberg to Houston to finally meet Armstrong, Aldrin, and Collins for some practice.

NASA had purchased an old Navy amphibious landing craft to support the training of astronauts in Galveston Bay, south of Houston. They named the craft "Retriever," and commonly used her to transport personnel into the bay for training and testing of equipment and materials associated with mock CMs. Clancy boarded Retriever to participate in BIG training with the Apollo 11 astronauts, but he had yet to meet any of them. He was leaning against a railing when he was joined by another man who asked whether he, too, was there for some training. Hatleberg said yes, and informed the newcomer that he was with UDT.

The mention of UDT ignited an animated conversation between the two men, as Hatleberg's new companion opined that it would be so cool to be a Navy commando. Hatleberg talked about his experiences with the teams and grew so comfortable during the conversation that he shared a confidence: Though it was indeed cool to be a Navy frogman, and he loved it, his secret desire was to someday be an astronaut, like the guys going to the moon. It was only after the training actually commenced that Hatleberg realized the person he had been speaking with was in fact Michael Collins, the CM pilot for Apollo 11. That conversation remained Clancy's most vivid and favorite memory of the entire recovery evolution.

The training went smoothly, with no glitches, but another challenge awaited Clancy on his return to Coronado to join his detachment. A three-man frogman team from UDT-12, led by Navy Lt. (jg) John McLachlan of Spokane, Washington, had joined the UDT Detachment Apollo unit, which was currently staffed by men from Clancy's parent UDT team, UDT-11. McLachlan's teammates were fellow UDT-12 frogmen PH2 Terry A. "Muehl" Muehlenbach and ADJ3 Mitchell L. "Buck" Bucklew. While they were veteran UDT swimmers who had previously received specialized training in Apollo recoveries, the change posed a real challenge for Hatleberg. With the flight of Apollo 11 less than eight weeks away, he had little time to get the UDT-12 team up to speed.

UDT Detachment Apollo still consisted of twelve men. In addition to Clancy, they included McLachlan's team, which was designated throughout the Apollo recovery operation by the helicopter call sign Swim ONE. Chesser's team, including Mallory and Wolfram, was designated as Swim TWO. A third team, led by team leader Lt. (jg) Robert R. Rohrbach, and team members ADJ3 Joseph "Joe" Via, and Joe's underwater demolition training (UDTRA) classmate, GMG3 Charlie A. Free, was designated Swim THREE. Free was the only member of the detachment who was a native of Coronado, California, and who had grown up within an environment dominated by the Navy special warfare ethos. EM2 Michel P. Bennett served as an alternate for Clancy Hatleberg in the role of decontamination swimmer, and HM1 Thomas G. "Doc" Holmes was an alternate swimmer who was qualified to substitute on any of the swim teams.

As for Hatleberg, he was more than ready for the splashdown of Apollo 11. He looked forward to once again seeing astronauts Neil Armstrong, Buzz Aldrin, and his new friend Mike Collins. They would meet again during the early morning hours of 24 July 1969, on the day of the *big* one.

9

En Route to Recovery

THE swimmers of UDT Detachment Apollo were among the last to board *Hornet* before she cast away lines to depart Pearl Harbor. They boarded at 0800 and were so tired from partying in Hawaii that they immediately hit their racks for a rest. While Hatleberg had split the workload among his three swimmer teams during the 7–9 July underway period off the coast of Hawaii, the bulk of practice recoveries for the next couple of weeks, including three of the five scheduled full-scale simulated exercises, or SIMEXs, would be performed by his primary recovery team, Swim TWO.

At 0903 that morning, *Hornet* slipped her mooring lines and was under way for the Mid-Pacific Line, and her launch abort station. It was Saturday. The launch of Apollo 11 was scheduled for the following Wednesday, 16 July 1969.

After launch the astronauts of Apollo 11 would orbit the earth to check the status of their equipment and instruments after the jarring of launch, and then begin to make their way to the moon. If their checks established that critical equipment or instruments had been irreparably damaged, then NASA would abort the entire mission, and Apollo 11's astronauts would prepare their command module (CM) for an early return to earth, later on launch day. NASA had already calculated the spacecraft's re-entry and splashdown points in the event a decision was made to abort the mission, which differed from the re-entry and splashdown locations designated for its regularly scheduled return from the moon on 24 July 1969; so *Hornet*'s first assignment after leaving Pearl Harbor on Saturday, 12 July 1969, was to take up station in what NASA designated as her launch abort station along the Mid-Pacific Line.

No recovery practices were held while the ship was en route to her launch abort station. Practices would have required the ship to slow her speed and remain in one general area until practices were completed. *Hornet* could not afford the extra time required to complete recovery or retrieval practice, so she simply gave her crews a rest. They would need to be fresh, in any event, if a mission abort was ordered on launch day.

General briefings were held for members of the Hornet Module Retrieval Team. And a press briefing was conducted at 1600 that day. During the press conference, the ship's navigator and her meteorologist issued routine reports about the ship's current location and the weather. Comments were additionally made about weather conditions predicted for the launch abort area, in case members of the recovery and retrieval teams had to perform operations to recover an aborted Apollo 11 mission a few days hence. Captain Seiberlich was available for almost all press briefings, and he routinely introduced guest speakers who included, on occasion, the commanding officer of HS-4, Commander Jones; the NASA team leader, Dr. Donald C. Stullken; the ship's first lieutenant, Lt. Cdr. Richard I. Knapp; the NASA engineer who operated and maintained systems in the mobile quarantine facilities (MQF) in hangar bay #2; the NASA medical group leader, Dr. William Raymond "Bill" Carpentier, who would examine the astronauts after splashdown, and then remain with them inside the MQF until their return to Houston; Lt. Clancy Hatleberg, who commented occasionally about UDT operations; the ship's air officer, Cdr. Irwin "George" Patch Jr.; and miscellaneous NASA technicians.

The Hornet Module Retrieval Team was informed that the Apollo 11 command module would experience a blackout period in their communications for up to four minutes before splashdown. If the CM came down more than twenty miles from the *Hornet*, and the ship's communications link with the CM was therefore lost, the air boss, aloft in an

E-1B aircraft and thereby likely to remain in communications with the CM, would take charge of the recovery operation and make all decisions regarding the operation. The CM was expected to contact *Hornet* by radio approximately five minutes before splashdown, as her main parachutes opened—her call sign was "Apollo 11."

Lt. Shirley H. "Lee" Elliott was scheduled to be on watch in the ship's combat information center (CIC) during splashdown. Elliott would initiate the ship's calls to the CM, and would therefore be the first person on earth to communicate with the astronauts after their fiery re-entry into the earth's atmosphere. Throughout the period of the expected communications blackout, the world would hold its breath, wondering whether the astronauts had indeed survived the heat of re-entry, and whether their parachutes had opened as designed to prevent them from free-falling to a sudden death as they slammed into the surface of the ocean. Lee Elliott would be the first person on earth to learn their fate as he called their call sign, Apollo 11, after the communications blackout, and awaited their response. Once that initial contact between *Hornet*'s CIC and the CM was accomplished, Lt. Elliott would pass the communications to the air boss, who was aloft. The air boss would then query the astronauts via radio about their status.

After the press conference, reporters rushed to interview the UDT frogmen. Reporters understood that, although Lt. Elliott would be the first person on the planet to speak with the astronauts after they returned to earth, the global public would be tuned in to the show because they wanted to actually see these men who had first walked upon the moon. In fact everyone wanted to actually see these men. And the first men on earth to experience first hand personal contact with Neil Armstrong, Buzz Aldrin, and Mike Collins after they returned from the moon were the swimmers of UDT Detachment Apollo.

The reporters' first question upon barging through the door of Ready Room One, which was designated for use by the frogmen, was "Who's the first swimmer in the water?" The swimmers were, at first, mystified by the question. What possible difference could that make? Initially, they had failed to consider that the first swimmer to enter the water would be the first human being on earth to see the astronauts after they returned from the moon. A demanding public wanted to know who that swimmer was and what he thought about his role in this historic enterprise. But the truth was, no one knew who that swimmer was going to be. The first UDT swimmer team to arrive on the scene after splashdown would deploy to install the flotation collar and assist the astronauts to egress their spacecraft, and that team's sea anchor swimmer would be the first in the water. But no one could predict which team might get to the CM first.

That was not the answer that reporters were looking for. They needed a body to interview. From their experience covering the recovery of Apollo 10, reporters knew that one of the UDT teams was designated, by their OIC anyway, as the primary recovery team. That team, in their estimation, was most likely going to get wet and accomplish the recovery operation.

That team was Swim TWO. The first swimmer to enter the water was most likely to be Wolfram, the youngest member of the detachment and the most junior in rank. At twenty years of age, Seaman Wolfram was also the fastest swimmer, and for that reason he was designated to serve as the sea anchor swimmer for Swim TWO—the first man in the water, who would attach a sea anchor to the CM as it sailed along in the wind.

When this became clear, reporters jumped at the chance to interview Wolfram. An article which appeared in the *Independent Press-Telegram* of Long Beach, California, stated, "Probably the most excited person connected with the recovery is 20-year-old frogman John Wolfram of (Fort) Atkinson, Wisconsin. He will be the first in the water and the first to touch the Module. 'It's really great! I'll be the first man here on earth to see 'em when they get back . . . the first one in the entire world'" ("*Hornet* Ready for Apollo 11 Recovery," *Independent Press-Telegram*, 12 July 1969, A-6). Ironically, had reporters not pointed that fact out to the swimmers, the thought might never have occurred to John Wolfram at all.

For the swimmers, media interest in them was completely unexpected, and it overwhelmed them a bit. They were excited about their own role in the recovery, but they did not believe for a minute that anyone else took notice. As far as they were concerned, they were on board *Hornet* to do a job and then go home. That was it. But Mike Mallory had a good time with the reporters. He hooked up with some of the reporters he had met during the May 1969 recovery of Apollo 10. In his diary, he wrote, "Hope it all works out so I am back in the water again this trip. I sure would be unhappy if I did not make the primary team on this, the BIG shot! (I will) play it by ear. Worry will get me nothing but trouble." The reporters seemed to be focused on Wes Chesser's Swim TWO, but the swimmers knew better. Even Mike Mallory, who was simply bigger than life and a member of Swim TWO, had his doubts.

USS *Hornet* steamed due west out of Pearl Harbor, before finally turning southward en route to her next station. Lt. (jg) Beaulieu was a young officer on the bridge of the ship as the sun set that day. He recalled that the sun was directly ahead of him on the horizon, blinding his eyes. The whole world then seemed to turn like a turntable as *Hornet* altered her course southbound toward the southern Pacific and the launch abort station, and the sun seemingly moved slowly from his front, facing him, to the starboard side of the ship as it sank into the horizon. It was an awesome feeling. And that night, when the brilliance of the moon was out, the moon which served as the target for the mission of Apollo 11, it obscured most of the stars. That was when the significance of their mission struck home.

SUNDAY, 13 JULY 1969

USS *Hornet* was a working vessel of the U.S. Navy and as such she was always on call to perform her mission. While at sea, men performed their jobs twenty-four hours per day, seven days per week. The work never ended. But even U.S. Navy ships provided distractions and an occasional amenity for members of the crew. Sunday newspapers were a favorite.

AC1 Anse E. Windham served as a supervisor in the Carrier Air Traffic Control Center (CATCC), where he functioned essentially as an air traffic controller. He devised a list of twenty frequencies which would be used by all aircraft involved in the recovery of Apollo 11, so all would be able to communicate throughout the operation.

Windham's boss, Lt. Ernie E. Connor, treated him well, and allowed Windham to fly as a passenger on the Carrier Onboard Delivery (COD) aircraft occasionally to pick up newspapers for the ship. It gave Windham a chance to get away from the routine of life on board ship and to perform, in the process, a worthwhile service for the crew.

Hornet launched one or two COD flights per day while operating at sea. The Northrop Grumman–manufactured COD aircraft, first introduced to the fleet just three years earlier, in 1966, transported both cargo and passengers between the ship and the nearest shore station; it had a maximum range of about 1,500 miles. The COD might fly to Pago Pago, for example, to pick up spare parts which had been shipped from Pearl Harbor for delivery to *Hornet*, mail, sundries, and crewmen who were returning from emergency leave, and Windham's newspapers. *Hornet* was a city which never slept and, like any city, her population needed to interact with their surroundings for logistical support, even as the ship steamed hundreds of miles away in the middle of the ocean. The carrier's COD aircraft enabled them to do so.

On Sunday, 13 July 1969, the ship's chaplain provided money for Sunday newspapers, and Windham flew in the COD to Pago Pago. The COD remained on the island for a few hours, and Windham passed the time drinking beer in the lounge at the Intercontinental Hotel. But he did remember to pick up a load of newspapers before returning to the ship, which continued on its southbound course toward the launch abort station.

On board *Hornet*, recovery and retrieval practices may have been suspended, but that did not mean the men were idle. BM2 Mickey N. Lowe was a member of the Hornet Module Retrieval Team and that day his assignment was to replace an attaching bolt in the housing suspended at the end of the in-haul line on the B&A crane.

Lowe recalled that NASA seemed to have an awful lot of money available to support recovery and retrieval operations, and they always did things first-class. They replaced nylon in-haul lines after every practice, for example. Given his experience in the Navy, that seemed an extravagant expenditure of taxpayers' dollars. But the cost of the 1.5-inch heavy nylon Sampson Braid line which was threaded through the B&A crane was absolutely nothing compared with the cost of the attaching bolt which Lowe was tasked to replace.

To get to the housing, Lowe had to crawl under and inside the B&A crane itself while suspended over the passing ocean below. He carried that special bolt in one hand. He had a lifeline attached to his waist, so he was not worried about falling overboard, but he was indeed concerned about

the possibility that he might accidentally drop that bolt. But he was successful. After he completed that task, a NASA technician told him that
it was a good thing he had not dropped the bolt because it cost $40,000.
At that time, Lowe thought he might have been taking home $200 per
paycheck, or $400 per month, so $40,000 sounded like an awful lot of
money to be tied up in one bolt for Apollo.

Lt. Cdr. William G. Tasker, the ship's meteorologist, participated in
the 0800 press conference that day. The weather was always briefed during the daily press conferences. Weather forecasts were actually provided
to him by the Manned Spacecraft Center in Houston, Texas, and they
were generally accurate. Tasker simply parroted what Houston related to
him every day.

Members of the press were always interested in weather reports,
because they assumed that rough weather would be a safety factor during
recovery operations. That was generally true, but Lt. Clancy Hatleberg
once confided to Tasker, during a conversation held in the wardroom
of the *Hornet*, that the weather was not really a significant factor for
his UDT swimmers. In fact they actually preferred a choppy sea state
because the wave action made it easier to install the flotation collar on
the CM.

Hatleberg's men spent that day testing the operational collars to ensure
they would hold air. The swimmers had brought on board *Hornet* two
different categories of flotation collars and stacked them atop one another
in *Hornet*'s hangar bay: those which they used for practice sessions—or
training collars—and those used exclusively for actual recoveries, called
operational collars. Swimmers could distinguish one from the other by
checking the serial numbers, which included a T for training collars, and
an O for operational collars. Operational collars were kept in pristine
condition to ensure they remained inflated during actual recoveries. After
completing the checks, the men packed the collars and all of their rafts.
It took all day, and they finished in time for dinner, around 1830. Mallory
summed up the day's work in his diary that night: "Boy, what a lot of
work." Shortly after chow, the swimmers hit the rack. Within hours, they
would be awake and preparing for a late night exercise.

While the swimmers worked in the hangar bay, Windham's newspapers were delivered throughout the ship. An article in that day's papers

about *Hornet's* historic mission commented on the amount of philatelic mail—some 250,000 envelopes—that crewmen expected to process on Splashdown day, the so-called First Day covers: "The *Hornet* will probably cancel more letters the day of the Apollo 11 splashdown than have ever been cancelled aboard a ship on a single day, in history."

A July 1969 *Los Angeles Times* article by reporter Charles Hillinger quoted *Hornet* postal clerk Richard L. Richards's comments concerning the overwhelming task of processing so many First Day covers on Splashdown day: "It sure would be a lot easier if we could start cancelling envelopes now, but we have to wait until splash down before we can put postmarks on the envelopes."

Members of *Hornet's* crew added to the list of people who requested the special stamps. Ens. Thomas M. Meisenhelder, a Navy Reserve officer who reported for duty on 3 June—just two days before USS *Hornet* was nominated to serve as PRS for Apollo 11—recalled that his father was among the many stamp collectors who wanted First Day covers, so he ensured that his dad received one.

Postal clerk Richards and his team planned to go topside on the day of the recovery to see the astronauts and to catch a glimpse of President Nixon, but then go down belowdecks to the library to begin processing the envelopes. Among those self-addressed envelopes sent in for the special First Day cover were six envelopes signed on the outside by Neil A. Armstrong, Edwin A. Aldrin Jr, and Michael Collins.

Equally important news was delivered by telegram that day to one of *Hornet's* enlisted men, Bruce Erickson: His daughter was born that day, 13 July 1969, exactly one week before Apollo 11 astronaut Neil A. Armstrong would become the first man in history to leave the gravitational confines of earth to walk on another celestial body in the universe, the moon. Erickson missed his daughter's birth. But he would definitely have a front row seat for the *biggest* show on earth.

The weather turned nasty while the UDT swimmers slept. Lt. (jg) Beaulieu recalled that he went down into the hangar bay alone that night, before the SIMEX, which was scheduled to begin at 0200 the next morning. It was raining, the ship literally shuddered as it plowed through the waves, and the wind blew weather into the darkened bay. Red night lights in the hangar bay bathed the boilerplate in a reddish glow, and rain streamed lazily down its sides. Beaulieu's mind wandered back to his childhood. When he was twelve, he read a science fiction book by Richard Marsten titled *Rocket to Luna*, a story about the first manned flight to the moon. Published in 1953, the book projected the first man would step on the moon in 1975. As Beaulieu reminisced down below in the hangar bay, a serious discussion ensued in the commanding officer's stateroom among Cpt. Seiberlich; his XO, Chris Lamb; NASA's Dr. Donald Stullken; and UDT Detachment Apollo OIC, Lt. Clancy Hatleberg about the advisability of conducting the exercise at all, given the horrendous sea state and poor visibility conditions.

It was the darkest night that *Hornet* had seen since she left Long Beach, two and one-half weeks earlier. Cloud cover obscured the stars and the moon, and it was inky black outside. Waves crested at twelve to fourteen feet, huge monsters which threatened to toss the 9,000-pound boilerplate around like a leaf in a gale, and the wind blew at a steady twenty-five mph, with stronger gusts. It was a crazy night to go for a swim. The wave action alone promised injury and threatened death.

Chris Lamb counseled against it. "You don't have to jump off a cliff to know that it's dangerous," he said. Lamb had spent years at sea, and he understood the power of it. This was a serious matter. Men truly could lose their lives. It was arguably irresponsible to even think of conducting the exercise. But Dr. Stullken insisted they proceed as planned. He reasoned that the swimmers must practice in the worst of conditions in order to prove that they were ready for the worst-case scenario on Splashdown day. He stuck to that position and would not budge. The decision, however, was Captain Seiberlich's. He naturally asked for input from Clancy. Hatleberg was among the men who would suffer the storm if Seiberlich elected to proceed with the exercise as planned. He therefore had more at stake personally than any other man sitting at the table. If he thought it best to cancel the practice, then Seiberlich would weigh the young frogman's opinion heavily into his decision.

Seated nearby, the XO bit his lip. But Hatleberg did not hesitate. His best team was on deck—Swim TWO. They could handle it. So the captain agreed to proceed. After that night, Lamb judged the frogmen of UDT Detachment Apollo to be the bravest men he had ever known. Not because they braved the storm. But because, after having done so, they continued to answer the call.

MONDAY, 14 JULY 1969

Hornet was operating about four hundred miles from Johnston Island. The SIMEX was scheduled to run from 0200 to 0600, for four hours, in the middle of the night. The exercise simulated a "distant splashdown/ worst-case scenario": *Hornet* finds itself inexplicably out of position, far away from the CM when it splashes down, and therefore is forced to race to the scene at high speed. Lieutenant (jg) Beaulieu recalled that *Hornet* pushed herself to the limit of her ability, given the sea state, and made revolutions for twenty-four knots en route to the boilerplate's position. The weather was so bad that crewmen had to close the elevator doors to block the wind from entering the hangar bay. QM3 Michael Mallory, a member of Swim TWO, documented in his diary that the swimmers were up by 0030, ate chow, attended a briefing, and loaded the helicopters with flotation collars and rafts. They were aloft by 0200, to start the exercise.

After the running lights of *Hornet* disappeared over the horizon, the world beneath them turned completely black. There was no way to discern just how far above the surface they were running, because there was no surface to speak of. Everything was darkness. Their helo raced at top speed toward its target, and the noise from the engines, the vibrations, and the throbbing of its propellers deafened them as they stared into the void. But as they approached the boilerplate, the helo's spotlight lit the sea around it like a flashlight spotting an object in a darkened room, and at least they could see the thing racing along in the current beneath them. The helo slowed its advance, but did not stop to hover. It circled the boilerplate, which continued to run in the current and the wind beneath them, caught in the light from above, and began a slow approach from downwind, at an altitude which kept the craft just barely above the reach of the highest waves.

John Wolfram, the sea anchor swimmer, was the first to jump from their helicopter to begin the recovery operation. He required no guidance or signals from anyone else. From experience, he knew when to leap from the door. But he waited until he was right on top of the boilerplate before he jumped. "Good thing, too," Mallory observed in his diary later that day, "because we had 25 knot winds and the Command Module was really moving." Mallory watched with some alarm, though, as Wolfram struck the boilerplate bodily and bounced into the sea. He shook off the blow and managed to grasp the sea anchor ring mounted beneath the side crew hatch, and to deploy the sea anchor. The mock spacecraft slowed noticeably, though wave action continued to toss it around.

Mallory and Chesser were next, with the flotation collar suspended between them. Installation of the collar by Wolfram, Mallory, and Chesser did not go smoothly. "Had a little problem on hook up of one of the straps," Mallory documented, "it was fouled due to bad packing (guys are too much in a rush)." But that was the least of their problems.

The three swimmers from UDT-12, designated Swim ONE, had volunteered to play the roles of astronauts during the SIMEX. Hatleberg seated them in a raft tethered to the flotation collar. And that was when the sharks showed up—likely drawn by the spotlight and the throbbing of prop wash on the surface of the water. Something about that bright light shining from above like the sun, and the curious object surrounded by animals which, in their black wetsuits, looked alluringly like prey, drew them to the surface, despite the storm.

The swimmers rarely saw them first. They were too focused and busy with their work on the spacecraft to notice anything moving in the water around them. Besides, outside the rim of the spotlight's glow on the surface, everything was black. It was only in the immediate vicinity of the module itself that anything could be seen in the water at all.

The helicopter crewmen usually spotted them first. From their elevated position hovering above the action down below, they could see huge gray shapes swimming beneath the surface, surrounding and circling the swimmers and slipping in and out of the light. A signal had been devised to enable helicopter crewmen to warn the frogmen when sharks were spotted. It was a simple hand signal. A crewman would lean out of the helo door and clap his arms together in an up-and-down motion to mimic the jaws of a shark opening and closing. Had it been daylight, a swimmer might have glanced up to see it. But at night, it was impossible.

The swim team leader, Wes Chesser, carried a radio in a DayGlo bag affixed to his waist, but he almost always shut the thing off to avoid distractions during the recovery operation. There was nothing worse than an interruption of static, followed by an inane question or comment from the helo hovering above when you were in the midst of trying to accomplish something down below. Nothing worse, that is, except sharks.

As Clancy seated the last of his three "astronauts" into raft #2, Mallory spotted the first of the sharks. "All three ('astronauts') had just gotten into the raft and were about to be picked up when I saw 4 sharks," Mallory wrote into his diary later that day. "I'm always checking for the damn things and tonight it paid off. Two of them were about four feet from John Wolfram, and two were about 10 feet from me. I usually have a very deep voice—but I came straight up out of the water and in the highest soprano voice was yelling shark for all to hear. . . . Boy, once we knew there were sharks in the water we were out in a second—and it was assholes and fins getting into the raft. When it came time for the pickup, the sharks were all over the pickup net. Boy, what a night."

AWHC Stanley C. Robnett was the hoist operator for HS-4's Recovery, and one of the crewmen who commonly saw sharks in the water circling the swimmers. He laughed as he recalled, "Those boys could come out of the water pretty fast," but the sharks were just getting started. Robnett lowered the Billy Pugh net to hoist the mock astronauts into his Recovery. Dangling beneath the Billy Pugh was a long nylon rope affixed to a small sea anchor. Its purpose was to provide drag, to keep the Billy Pugh net from flying around in the prop wash once it hit the surface. A shark had that sea anchor for dinner. Mallory recalled, "There is a small bag on a nylon rope, it contains a small sea anchor. Well, some shark has its own sea anchor now!"

Mitch Bucklew was a member of Swim ONE who played the role of astronaut for that SIMEX. Bucklew was the only member of UDT Detachment Apollo who served during his ten years of Navy service with both UDT/frogman and Navy SEAL teams. In fact, Bucklew was a member of the very first cadre class of Navy SEALs. During his stint in the Navy, Bucklew deployed four times on combat tours in Vietnam; once with SEAL Team ONE, and three times with UDT frogman teams. He was one of only a handful of Navy UDT/SEALs in the history of that organization who completed "Hell Week" on three separate occasions, with UDTRA (basic UDT training) classes 30, 35, and 37, owing to injuries which set him back twice during training. Bags of a World War II–era shark repellent—copper sulphate—were suspended from the Billy Pugh net on which he was supposed to ride during hoisting operations into the hovering helicopter. But it did not work as advertised that night. "The sharks just ate them," Bucklew later recalled.

As *Hornet* approached the boilerplate and swimmers to begin the retrieval operation, the frogmen ignited flares so their position could be seen by men on the ship. Their effort was in vain. According to Mallory's personal diary, "The rest of the night turned out to be very long. The ship goofed up on their first pass and had to make another one. And did he ever take his time. Meanwhile, we bobbed up and down until the sun came up."

To everyone's relief, the second pass was perfect. *Hornet* passed to within sixty feet of the boilerplate. And Mallory was not taking any chances that the ship would mess up this time. NASA's Apollo Recovery Operational Procedures Manual directs swimmers to protect themselves from injury when the shot line is fired by hiding behind the module itself. But whoever wrote that procedure obviously did not know Mike Mallory. He stood front and center on the flotation collar and made a huge circle of his arms, a target, and challenged the gunner's mate with the shot line to give it a go. The gunner's mate fired the shot dead center, and from that point on, Mallory always made a target of himself.

Even then the sharks did not let up. Beaulieu participated in the operation to retrieve the boilerplate from the sea after completion of the SIMEX. He recorded in his notebook that seven to ten sharks were spotted nearby during the retrieval operation. Naturally, the swimmers were still on and near the boilerplate throughout the operation. Beaulieu realized that the dumping of garbage off the fantail approximately thirty minutes before sunset and after sunrise attracted the sharks. And since the retrieval exercise occurred near dawn, *Hornet*'s routine was putting the UDT swimmers in danger.

FA William C. Foss had just joined the Navy and reported for duty on board *Hornet* in May 1969, as she completed her final WestPac cruise and pulled in to her home port in Long Beach, California. He was just a seaman apprentice (E-2) at the time, and one of his "duties," as one of the lowest-ranking men on the ship, was to dump garbage off the fantail. He had not imagined that, in accomplishing such a mundane duty, he was actually chumming for sharks to attack UDT swimmers and potentially threaten the very success of the Apollo 11 mission. He was happy to stop the practice.

At the end of the SIMEX, Chesser's UDT team assisted with the retrieval operation and then prepared to board ship as only UDT frogmen might. The ship tossed a cargo net over the side, aft of the B&A crane, and the swimmers climbed the net and up onto the ship. Climbing the net might have been a routine action, but for the scullery discharge. The ship's kitchen routinely dumped waste water, which contained a high concentration of food particles, through lines which flowed directly into the sea. The sudden discharge, which occurred while Swim TWO climbed the cargo net, threatened to knock the swimmers back into the sea. To make matters worse, the discharge attracted more shark activity. Beaulieu made another note: Locate and close scullery discharge during retrieval operations. The crew still had a lot to learn about Apollo recoveries.

It had been a long night. Or rather, a long morning. The sun had just risen above the horizon, and the work of cleaning and storing equipment after the SIMEX was not yet done. But it had been a good practice, a tough practice under the most challenging conditions, and the Hornet Module Retrieval Team felt more confident than ever that they would acquit themselves well when it came time to retrieve the Apollo 11 command module from the ocean; the module, and its precious cargo of moon rocks. *Hornet's* crew never lost sight of the fact that their efforts would yield the first samples of soil ever returned from the moon.

The Soviet Union developed its manned spaceflight program behind a heavy curtain of secrecy. And people in the West could only surmise that the United States and the Soviet Union were in a race, of sorts, to place the first man on the moon, given the quality of early Soviet successes in space operations—the first satellite in orbit, Sputnik; the first dog in space; the first man to orbit the earth; the first space walk, or extra vehicular activity (EVA); and the first landing on the moon of an unmanned vehicle in 1966, and so forth.

But the Soviet Union had secretly stopped running the race. For reasons which are still not clear—though possibly attributable at least in part to repeated failures by the Soviet Union to develop a dependable rocket comparable to America's Saturn V rocket—the Soviet Union launched a parallel effort to develop not manned exploration of space, but an unmanned probe, a robot, which could perform at least some of the same duties as a man on the moon.

Their progress in developing such an instrument was remarkable, given the relatively primitive level of technology available at the time. It was a solar-powered remotely piloted lunar rover which was designed to land on the moon, scoop soil samples, and then return the samples to earth. NASA had nothing like it. The Soviets called it Lunokhod, or Lunar Rover, and they launched it as Luna 15 on Monday, 14 July 1969, two days before the launch of Apollo 11. At the time, the launch of Luna 15 was viewed as an effort by the Soviet Union to obtain lunar soil samples and return them to earth before American astronauts would do so. The technology developed to support Lunokhod—solar power, and the ability to remotely pilot the robot from earth—was not duplicated by NASA until 1996, some twenty-seven years later, in the Mars Pathfinder craft.

The launching of Luna 15 was a shock. And it took some of the wind out of NASA's sails. Dr. Stullken, the leader of NASA's recovery team on board USS *Hornet* on the day Luna 15 was launched, put the Russian effort into perspective: "The Russians will steal a lot of our thunder if they get moon samples back before we do. It will be a great feat to have a man get out and walk on the surface of the moon, but a number of scientists are actually more interested in obtaining surface (soil) samples for analysis (than landing a man on the moon)" ("Russ to Steal Apollo Thunder if Feat Works," *The Daily Plainsman,* 14 July 1969). Despite the news, *Hornet* pressed ahead, still en route to her launch abort station. There was nothing the crew of USS *Hornet* could do about seeming Russian skullduggery. NASA would have to deal with it. *Hornet* had her own challenges, nearer to hand.

TUESDAY 15 JULY 1969

Apollo 11 was scheduled to launch the next day. *Hornet* continued to steam southbound into the Pacific Ocean to reach her launch abort position, just in case system malfunctions forced astronauts Armstrong, Aldrin, and Collins to terminate their flight prematurely and return to earth shortly after their launch.

Most of the UDT swimmers had earned their Shellback certificates during an earlier voyage. So they had nothing to worry about that day; in fact, the raucous Shellback Ceremony was something they looked forward to. They celebrated the dawn of its arrival early that morning in a typically Navy frogman way, with a run.

At sunrise, the hard-charging commandos shattered the relative tranquility of an otherwise peaceful Navy morning with an exuberating round of physical training (PT), frogman style. All over the ship. It was loud, full of energy, and smacked of a boisterous camaraderie which was intended to impress upon lesser mortals on board *Hornet* that UDT frogmen were perhaps, as they were reputed to be, a somewhat different breed of men.

They ran in formation up and down ladders, then up to and around the flight deck. They loved this stuff—lived for it, in fact. The swimmers were used to constant physical challenges and the freedom granted by wide open spaces to vent their energy. They had been cooped up for days within the confined space of a Navy ship, and, like thoroughbreds, they simply needed a chance to run in order to feel right. "We really shook up the ship," Mallory, the workhorse of Swim TWO, recounted in his diary. "They could not believe us. Even the big bad Marines got a shock!"

The press release from *Hornet* for 15 July 1969 looked beyond the launch and the possibility of a prematurely terminated flight to the moon, to the projected recovery itself. Reporters interviewed Clancy Hatleberg, the twenty-five-year-old frogman from Chippewa Falls, Wisconsin, whose role as BIG swimmer during the recovery operation meant that he would be the first person on earth "to come within touching distance of the Apollo 11 astronauts [after they returned to earth]."

The reporters queried Clancy about moon germs and any concerns which he entertained about contracting a plague through contact with the astronauts. "I'm not worried one bit about myself," Hatleberg assured them. He expressed confidence in NASA's quarantine system and the procedures developed to protect earth from back contamination. "Even if they do bring back something," the young swimmer declared, "the rest of us will be protected. The NASA engineers, technicians and scientists have done a topnotch job making sure we will have no problems in this area." When asked to characterize the difficulty of a recovery operation itself, Hatleberg responded, "It's a piece of cake provided you have good weather."

10

Launch

THE three major TV networks in the United States, ABC, NBC, and CBS, each devoted more than thirty hours of continuous broadcast coverage of the 16 July 1969 launch of Apollo 11. Their estimate of 500 million viewers, some of whom watched television coverage round the clock prior to the launch, testified to the depth and breadth of interest in humankind's first attempt to place a man on another celestial body, the moon.

NASA's manned spaceflight program had its birth just eight years earlier on 25 May 1961, when President John F. Kennedy pronounced a national goal "before this decade is out, of landing a man on the moon and returning him safely to Earth." The cost of NASA's effort to realize that goal was approximately $24 billion. That seemed an inordinately large investment to some, at a time when the average income in the United States was just $8,500 per year, when a new house could be purchased for just $16,000 and a new car for less than $3,500. Tuition to Harvard University in 1969 was only $2,000 per year. Milk was $1.26 per gallon, a loaf of bread 23¢, and a gallon of gas just 35¢. But the relative cost of the program was actually quite small. It amounted to approximately one-third the amount of money Americans spent on cigarettes during that time frame or approximately one-sixth of what Americans spent on alcoholic drinks.

It seemed that everyone was talking about the flight of Apollo 11, and a number of men exploited the broad interest in it to their own benefit. A Greek insurance underwriting firm, for example, wrote an "astro policy" to insure the lives of the Apollo 11 astronauts for $10,000 each. It was a publicity stunt. The company observed, however, tongue in cheek, that the policies would be rendered void if the astronauts were kidnapped by extraterrestrials or if they found the moon to be so attractive that they refused to leave.

For all the hoopla around the world which accompanied the launch of Apollo 11 into space that day, life on board USS *Hornet* remained essentially the same. It was business as usual. She stood by to assist in the event astronauts Armstrong, Aldrin, and Collins aborted the mission shortly after launch, owing to failure of critical systems on board their craft. Their command module, *Columbia*, named after Columbiad, the cannon used to launch the spaceship in Jules Verne's acclaimed 1867 novel *From the Earth to the Moon*, would then re-enter the earth's atmosphere and splash down somewhere in the vicinity of *Hornet*'s launch abort position, and that would be the end of that.

TVs on board *Hornet* did not provide coverage of the launch. It was technically impossible to do so at the time. When informed that the launch had been successful and that the flight of Apollo 11 would proceed to the moon as planned, *Hornet*'s crew simply went back to work after they received some motivational words from their commanding officer. Cpt. Carl Seiberlich addressed his crew via the 110 closed-circuit televisions on board *Hornet*. He understood that the die had been cast. Men were truly on their way to the moon. And within just a few days, the entire world would watch as *Hornet* executed the final and arguably most dangerous phase of the flight: splashdown and recovery of the astronauts. Seiberlich declared in his message to the crew: "The three most important people in the world at this time in history will come aboard our ship. We are the last link in a very important chain." That operation just had to succeed, without a hitch.

They wasted no time in making their preparations. The crew conducted a boilerplate exercise during daylight hours that day, from approximately 1000 to 1230, utilizing the B&A crane equipped with a 6-foot nylon strap. The weather remained a concern, and Dr. Stullken, chief of NASA's onboard recovery team, feared that the boilerplate was rising and dropping in such heavy seas that it might hit the bottom of the elevator. He directed the crew to switch from the B&A crane to the Tilley crane if the weather worsened, to avoid such damage. As the seas increased in height, the crew switched from a 6-foot nylon strap at the end of the crane's cable, to an 8-foot strap, and finally to the longest strap provided by NASA for retrieval operations, a 12-footer.

Dr. Stullken noted that rotor wash from helicopters hovering above the boilerplate tended to "push" moon germs toward the UDT raft tethered upwind from the boilerplate. He observed during practice that day that one of the UDT swimmers briefly removed his mask and thereby possibly inhaled some nasty moon pathogens. Stullken reminded them that swimmers must not remove their masks, and that they must remain on Scuba to breathe, until the astronauts and the side crew hatch of the spacecraft were sanitized by the BIG swimmer. Failure to remain on Scuba throughout that evolution would constitute a violation of the decontamination procedure and force NASA to quarantine the swimmers inside an MQF after the recovery operation was completed. NASA did not want to see the swimmers breathing moon germs out there in the middle of the ocean.

NASA's procedures seemed very rational on paper and in theory, and were even practical, perhaps, when executed under ideal conditions—when performed at a leisurely pace in a swimming pool in NASA's facility in Houston, Texas, for example. But the open ocean was another matter.

Mike Mallory was one of the swimmers assigned to UDT Demolition Apollo's primary recovery team, Swim TWO. Mallory remarked that installation of the flotation collar in a choppy sea was hard work. It took a lot of air. And although the swimmers brought extra scuba tanks along for the exercise, which were stored in the ship's whaleboat until needed, the swimmers usually used up all of the air in their scuba tanks long before the recovery operation was completed. So they cheated, just to keep Dr. Stullken off their backs. After depleting their scuba tanks of air, the frogmen would simply bite down on their mouthpieces and pretend to draw air from their tanks while actually breathing out of the sides of their mouths. Moon germs be damned. Mallory documented an evaluation of Swim TWO's practice that day: "It went just fine—which made everybody very happy." But Mallory and teammate John Wolfram then dragged every scuba tank in UDT's inventory, half the length of the ship, to be recharged after the practice session.

Lt. (jg) Beaulieu spotted two more sharks following in *Hornet*'s wake during the boilerplate retrieval operation, and made a note to ensure that extra men were placed on shark watch, in addition to the armed sailor on board the whaleboat. GMG3 William W. "Bill" Winchell, a rancher from Montana and a veteran of riverboat patrols during the Vietnam War, led a team of armed gunner's mates who were posted all around the ship to watch for sharks. After retrieving the boilerplate from the sea, *Hornet*'s crew practiced moving it from the elevator to the mobile quarantine facility (MQF) down below, in hangar bay #2.

During the press conference on board *Hornet* later that afternoon, Stullken expressed his confidence in the Hornet Module Retrieval Team: "We are in a horse race out here. At this stage of the game the Hornet is ahead of any ship we've had since the beginning of the space program" ("Hornet Prepared as Final Link in Chain," *San Antonio Light*, 16 July 1969). Stullken had been working hard, too. Lieutenant (jg) Beaulieu, who commanded one of the deck crews on the retrieval team, recalled that Stullken and his assistant, John C. Stonesifer, worked nearly nonstop for two weeks, attending to every detail of the recovery and pushing the crew for perfection to ensure a successful mission. Stullken wore a T-shirt throughout that time, rather than the standard NASA white dress shirt, as a message to *Hornet*'s crew that he was on board to work, not on some vacation.

Hornet's crew was focused, and now fully familiar with Apollo operational recovery and retrieval procedures. They had only to continue to practice and hone their skills. Little setbacks continued to plague the team, however. Lt. Cdr. Joseph A. Fidd, *Hornet*'s supply officer, recalled that the Hornet Module Retrieval Team lost one of three HT-200 Motorola walkie-talkies during the practice session. The walkie-talkies had been purchased especially for the retrieval operation, to support communications between the bridge, the B&A crane, and the whaleboat. The loss of just one of those walkie-talkies became a single point of failure for the mission—poor communications produced poor coordination, which, in turn, could result in failure of the mission. Fortunately Fidd had transferred from the naval support center in Long Beach, California, then joined *Hornet* in Honolulu just a couple of weeks earlier so he still knew the supply officers back in Long Beach. They could be counted on to expedite an emergency request for support.

Apollo recoveries were a big priority for the Navy, anyway. Requisitions for supplies which supported such operations were given their own Apollo Project supply code of 708. The application of that code on a requisition form ensured expeditious handling of the order. To ensure that the system worked smoothly, *Hornet* detailed one of its storekeepers to the naval supply center in Pearl Harbor and another to Johnston Island, which lay in relatively close proximity to *Hornet*'s splashdown position, to coordinate and facilitate supply requests. Special routing cards were affixed to items ordered to support Apollo recoveries. In this particular case, Fidd's buddies at the naval supply center in Long Beach filled the order, then sent a message to Fidd and storekeepers in Pearl Harbor and Johnston Island, which detailed the supply flights that carried the extra walkie-talkies Fidd had ordered. Fidd dispatched a COD flight from *Hornet* to retrieve the walkie-talkie replacements on their arrival at Johnston Island and to bring them back to *Hornet*—which still operated at sea—in less than twenty-four hours.

At night, crewmen of USS *Hornet* commonly ventured to the flight deck for some air. Flight operations were typically suspended during the night unless a practice session was in progress, so the flight deck was open and free for men to walk and talk, take in the night air, and look at the moon, if cloud coverage permitted. The night after Apollo 11 was launched on its journey to the moon, CWO2 Homer V. "Mo" Morris and some buddies were walking on the flight deck. The Mutual Broadcasting van was positioned on the starboard side of the carrier's island. As they passed that van, Morris and his buddies heard the sound of music emanating from within the van. They poked their heads inside the van and spoke with Mutual Broadcasting technician James Joseph O'Connor, who was listening to a radio station broadcasting from Hawaii. After hearing an advertisement on the radio for a pizza parlor located in Honolulu, Morris asked O'Connor whether his van was equipped to place a telephone call to Hawaii. It was. Morris placed the call and ordered a dozen pizzas for delivery to Pago Pago. The young man taking his order hesitated a moment, and expressed doubt about his ability to deliver the order. Morris was disappointed. He informed the young man that he was on board USS *Hornet*, operating somewhere in the vicinity of Pago Pago, and he hoped to receive delivery of the pizzas in time to celebrate the recovery of the Apollo 11 astronauts after they splashed down in the middle of the Pacific Ocean. There was going to be a big party on board ship afterward, and the crew was going to be hungry. When the kid stammered that he could not manage the delivery, Morris and his buddies doubled over in laughter.

But not everyone on board USS *Hornet* that evening was in a jovial mood. At 9:15 that night, the OOD on the bridge received an injury report from down below. A seaman apprentice assigned to the OI Division (within OD, responsible for tracking all surface and air contacts) was discovered on the floor of the head, with multiple cuts on his left wrist. Dr. Stephen A. Habener, the ship's surgeon, treated the young man and admitted him to sick bay for observation. Habener had served on board *Hornet* for only about a week by that time, having reported for duty after the ship entered port in Pearl Harbor on 10 July. His only patients since that date had been sailors seeking circumcisions, possibly to avoid hazing during the Crossing the Line ceremony. Now, this. The young man was Habener's first serious patient, and the young surgeon never forgot the incident. The sailor's wounds were not life-threatening. But on the day that millions celebrated the launch of the first of humankind's attempts to land a man on the moon, the young man had attempted suicide—or, at least, he wanted others to believe that he had done so.

THURSDAY, 17 JULY 1969

The astronauts of Apollo 11 were en route to the moon, a journey of approximately 240,000 miles. Their return to earth and splashdown into the Pacific Ocean was scheduled to occur in exactly one week. *Hornet* turned away from her launch abort position and made rotations for her splashdown position, farther south.

Recovery and retrieval exercises were held during a three-hour period from approximately 0820 to 1120 that morning, using an 8-foot nylon strap at the end of the B&A crane's cable. A rough sea state required the extra "give" of an 8-foot strap to avoid snapping the cable during the hoisting operation as the boilerplate cleared the water for retrieval from the sea.

At 1300, a presidential SIMEX commenced, and required but forty minutes to complete. The presidential SIMEX was a practice session, a walk-through to ensure that the ceremony to welcome President Nixon on board *Hornet* during the early morning hours of 24 July 1969, Splash-down day, would go smoothly. The president was scheduled to depart the White House in Washington, D.C., several days before splashdown, to begin an official presidential visit to various nations in East and South Asia and in Europe. Richard M. Nixon had been inaugurated president on 20 January 1969, just six months earlier, and the tour of nations in July and August 1969 marked his second official overseas visit as president of the United States. He wished to march through that tour in triumph, as president of the nation which had placed a man on the moon. Nixon's participation in the welcoming ceremony for the Apollo 11 astronauts on board USS *Hornet* after the recovery of Armstrong, Aldrin, and Collins from their splashdown in the Pacific Ocean, then, served as an exultant kickoff for his overseas trip.

Nixon was scheduled to fly first to San Francisco, to spend the night. Next morning, his entourage planned to fly Air Force One, his official presidential aircraft, directly to Johnston Island, in the Pacific, while First Lady Pat Nixon flew directly to Honolulu. Pat Nixon did not plan to participate in the welcoming ceremony for the Apollo 11 astronauts, possibly, in part, because Navy regulations denied women access to U.S. Navy warships. Accompanying Nixon at that point was his chief of staff, Bob Haldeman, who, much later in Nixon's presidency, would be indicted for his role in the Watergate scandal. The rest of the presidential party included America's secretary of state, Bill Rogers, who naturally accompanied the president in support of his diplomatic visits abroad; the national security adviser, Henry Kissinger; NASA director Dr. Thomas O. Paine; and Col. Frank Borman, who served as liaison between NASA and the White House for Apollo 11. Borman had served as the commander of Apollo 8, the first manned lunar orbit mission. The White House press corps also accompanied the presidential party.

From Johnston Island, the presidential party expected to fly by presidential helicopter, designated by call sign Marine One, to USS *Arlington* (AGMR-2), a major communications relay ship which was already at sea to support the Apollo 11 recovery effort. *Arlington* was built in 1944 on a heavy cruiser hull, but was first commissioned as the light aircraft carrier *Saipan* (CVL-48). She therefore had a flight deck which was capable of handling presidential helicopters. The ship was later converted to a major communications relay ship and re-commissioned as *Arlington*. The ship supported Frank Borman's Apollo 8 mission at the end of December 1968, in part by relaying then-President Johnson's congratulations to Borman's crew on a successful mission around the moon.

Nixon's party expected to spend the night on *Arlington* at sea, as she steamed toward the Apollo 11 recovery area. During the early morning hours next day, Splashdown day, Marine One would fly from *Arlington* to *Hornet* to participate in the welcoming ceremony for the returning Apollo 11 astronauts. The presidential party expected to remain on board *Hornet* for only a few hours to complete the ceremony, and then to depart in triumph to continue Nixon's second overseas trip as president of the United States.

The presidential helicopter fleet was operated by the U.S. Marine Corps, and though their pilots were experienced Marine Corps pilots, they had little hands-on experience navigating over open water at sea. There was justifiable concern, therefore, that Marine One might lose its way during the hop from Johnston Island to USS *Arlington* on 23 July 1969, or from *Arlington* to *Hornet* during the early morning hours of darkness on 24 July 1969. To ensure that Marine One pilots managed to stay on course and not get lost at sea with the presidential party on board, the Navy implemented a procedure called Operation Leap Frog. Operation Leap Frog simply required Marine One to employ its TACAN, or its tactical navigation equipment, to home in on electronic beacons which emitted from a series of U.S. Navy ships stationed along Marine One's flight path. The ships, in turn, tracked Marine One's progress via radar. Pilots of Marine One called every minute to each U.S. Navy ship positioned along her flight path to receive from air traffic controllers on board those ships their directions to correct Marine One's flight path. In that manner, Marine One was passed off from ship to ship, until she arrived at her destination, thereby preventing the American president from getting lost at sea.

Lt. Henry Francis Dronzek served as the administrative assistant to the commanding officer of USS *Hornet*, and was selected to serve as the master of ceremonies for the president's arrival. After Marine One's touchdown on the deck of the *Hornet*, sometime very shortly after 0500 on Splashdown day, *Hornet* gunner's mates under the command of GMG3 William W. "Bill" Winchell—the young enlisted rancher from Montana who claimed to have been "born on a saddle horse and with a gun in my hand"—would fire a twenty-one-gun salute (author interview of William W. Winchell, 10 March 2008). For the presidential SIMEX, *Hornet's* chaplain, Cdr. John A. Piirto, role-played the part of President Nixon. As the president deplaned from Marine One, Lt. (jg) James E. Ovard, the young officer who aspired to someday work as a broadcast journalist, would salute Nixon and officially welcome the president of the United States on board USS *Hornet*. Admiral McCain (CINCPAC), Rear Admiral Davis (TF-130), and Captain Seiberlich, representing USS *Hornet*, would then greet the president.

The president would be escorted into the ship's island and up to the flag plot, where he would be introduced to Lt. (jg) Richard F. Powers III. Powers was specifically selected by Captain Seiberlich to serve as USS Hornet's personal liaison to the president throughout his stay on the ship. Powers was responsible for ensuring that the president's every need was met.

Insofar as the president was expected to arrive on board *Hornet* nearly an hour before the calculated splashdown of Apollo 11, *Hornet's* crew prepared a breakfast tray for him. In typical Navy fashion, *Hornet* issued an official memorandum which not only detailed the items to be included on the presidential breakfast tray, but also identified every man who would participate in the preparation of that tray and choreographed the movements of every man expected to participate in the presentation of the tray to the president. The senior enlisted steward on board *Hornet*, SDCS Charles Everett Raigans, from Oklahoma, was given the honor to present the tray to the president. A card was placed on the tray which identified Raigans as the presidential server.

The next, and final, phase of the presidential SIMEX that day was the welcoming ceremony for the astronauts. While Nixon and his entourage remained in the flag bridge, the astronauts would be recovered from their spacecraft by UDT swimmers at sea and then transported to *Hornet* by the HS-4 Recovery helicopter #66, piloted by HS-4 commanding officer, Cdr. Donald S. Jones, of Madison, Wisconsin. Upon arrival, the band would play a song entitled, "Columbia, the Gem of the Ocean," in reference to the Apollo 11 command module, *Columbia*. The HS-4 Recovery would be lowered by elevator from the flight deck to the hangar bay below, and the astronauts, still garbed in their BIG suits, would then walk a short distance from Recovery to an MQF. Nixon and his entourage would remain on the flag bridge until the astronauts completed a shower and a brief medical examination. For the purpose of the presidential SIMEX that day, the ship's chief master at arms next escorted presidential role-player, Commander Piirto, along a preplanned presidential route from the flag plot to the hangar bay. Lieutenant Dronzek then called the crew to attention and ran through the presidential welcoming ceremony, music and all. Nothing was left to chance. The entire procedure was choreographed and practiced endlessly, until all participants mastered it by rote. By Splashdown day, the only participant in the ceremony who lacked practice would be the president himself. But *Hornet*'s crew had that covered as well. They intended to simply take him by the hand.

Everything seemed to be going according to plan. Members of the press pool scrambled to find interesting, if not sensational stories to report to an eager and expectant public. They honed in on the swimmers after the morning practice session and asked about sharks. They were not disappointed. Wesley Chesser, the leader of Swim TWO, related to them quite candidly after the morning practice in the water: "Sharks out there are swarming all over the place." His teammate, the young sea anchor man from Wisconsin and fastest swimmer in the detachment, John Wolfram, described the action for reporters: "They come right out of the water, three, four feet from us. We do not know what kind of sharks they are. They're long and gray. The sharks watch us. We watch them. They've been nibbling away at our gear all week" ("Splashdown Area Alive with Sharks," *Ogden Standard Examiner*, 18 July 1969, 7A).

UDT swimmers understood the media appetite for interesting tidbits about their training experiences, but the constant pestering by reporters for ever more sensational stories eventually grew tiresome. They conspired with their HS-4 teammates to teach the reporters a lesson. Pilot Richard J. Barrett recalled, "One night after practice as we were waiting to recover aboard, we were talking on the intercom and kind of complaining among ourselves about how the media would corner us at times trying to determine if we had any experiences during practice that might be newsworthy. We decided to tell them that a giant jellyfish attached itself to the sea anchor on the boilerplate. . . . I later heard that one 'prominent newspaper' actually ran that story, back on page 7. Commander Jones asked me if I knew anything about the origin of that 'rumor.' . . . I think I blamed it on John (Wolfram), the youngest member of the UDT detachment on board for the recovery of Apollo 11" (e-mail from Richard J. Barrett, 01 July 2008).

During the 1600 press conference, Captain Seiberlich said that he was "very pleased" with the morning recovery exercise. He noted that one purpose of daily recovery training was to establish ground rules for positioning the ship during splashdown "to avoid possible contamination of the ship," which was sensational to reporters looking for anything remotely interesting to send over the wire in the days leading up to the recovery. If the entire ship was contaminated by moon germs during the recovery operation, then everyone on board the ship, including the president, could be required to remain on board the ship throughout a twenty-one-day quarantine period. They would be stuck in the middle of the ocean, with the president on board.

Reporters pressed Dr. Stullken, the NASA team leader, to explain how contamination of the ship might occur, given the great distance the ship was expected to maintain from the command module (CM)—ten to fifteen miles—at time of splashdown. It seemed a logical deduction, to reporters, that *Hornet* would be out of harm's way, as it were, if it maintained that distance. Dr. Stullken dryly explained that air vents in the CM opened when the parachutes deployed during descent toward the surface of the ocean, and were not closed until after splashdown. Pathogens could escape from the inside of the CM through those vents, during the CM's descent to the surface. "If the ship was close enough (to the Command Module) downwind," Stullken conceded matter of factly, "it might be possible for the ship to be contaminated" ("Contamination

Possibility Remains Real as Hornet Readies for Recovery, Visitors," *The Dominion*, 22 July 1969, 3-A). His tone of voice left no doubt, however, that Stullken considered the possibility of contamination of *Hornet* by moon pathogens to be extremely low. To his mind, it was a non-issue. To drive a nail into the contamination coffin, Captain Seiberlich assured reporters that the ship would maneuver in such a way during the spacecraft's descent to the surface as to be positioned upwind from it, and therefore out of harm's way.

Members of the press corps, however, wanted more. They pressed Clancy Hatleberg for comments regarding the threat posed to swimmers and astronauts by shark bites. Reporters had received some good quotes on the subject from UDT swimmers Wes Chesser and John Wolfram that morning, and so the subject of sharks and shark attacks seemed like fertile ground for human interest stories. But Hatleberg, as if on cue from Captain Seiberlich, downplayed the threat when he judged quite confidently, "You have about as much chance of being hit by lightning" (USS *Hornet* message, 15 July 1969, Subj: Presrel—Daily Wrapup 15 Jul 69 [courtesy of Cpt. Chris W. Lamb, USN (ret.)]). That was the end of that story.

The rest of the 1600 press conference that day was banal at best. Lt. Cdr. Richard I. Knapp was a thirty-four-year old officer from Long Beach, California, who served as the ship's first lieutenant. In that capacity, he was in charge of members of the Hornet Module Retrieval Team who hoisted the spacecraft on board *Hornet* that morning with the B&A crane. Knapp explained to reporters that the winch crew used nylon line, which had a stretch factor of 50 to 60 percent, which allowed the recovery to be performed in heavy seas. But weather in the area had been good, with only two- to four-foot seas. A broken line and consequent loss of the CM, with its precious cargo of moon rocks on board—he assured the press—was not likely to be a problem.

In truth, the crew of *Hornet* did play some cards a bit close to the vest. During an informal discussion between NASA personnel on board ship and members of the crew, for example, *Hornet* officers learned that NASA had stored fifty-five one-gallon containers of sodium hypochlorite in a bulk container on the hangar deck. Storage of large quantities of that material in an open area like the hangar bay represented a serious threat to the ship, especially if heavy seas caused NASA's pallet to slide into anything which might puncture their container. *Hornet*'s officers took possession of the material and stowed it in an unused pyrotechnic magazine off hangar bay #2, immediately adjacent to the NASA working area. Talk of sharks with members of the press was okay, to a limited extent. But revelations about improperly stored hazardous materials, which might cause some to question the competence and judgment of NASA and Navy officials and which could endanger the entire crew and their civilian guests, as well as the astronauts themselves, was not. It was not mentioned to the press.

Hornet's engine trouble was also kept from reporters. Few crew members, beyond the commanding officers and those responsible for making repairs, were aware of the problems emanating from the engine room. MM1 Enoust E. Lee, the senior machinist's mate from Southside, Tennessee, who could fix just about anything, had reported a burned-out bearing on one of *Hornet*'s propeller shafts a week earlier. *Hornet* was operating off the coast of Hawaii at the time, practicing Apollo recovery and retrieval procedures. The ship pulled into port in Pearl Harbor on 9 July, and the bearing was replaced. But the stress of driving through a storm a few nights earlier, compounded by the additional stress of driving through medium seas since that date, must have proven too much for the bearing. On 17 July, the bearing seized up again, and one of *Hornet*'s propellers consequently fell out of action. It was, however, too late to return to Pearl Harbor for additional repairs. Apollo 11 had been launched the day before, and her schedule was therefore fixed. She would return to earth on Thursday, 24 July 1969, and splash down into the Pacific Ocean, whether *Hornet* was there to lend assistance to the astronauts or not. Machinist's Mate First Class Lee and his team were forced to make the necessary repairs at sea.

Hornet's engineering officer, Lt. Cdr. Robert P. Schmidt, and Machinist's Mate First Class Lee spent two days working on that bearing. They did so with but little sleep. Lee slept for perhaps four hours. The bearing was about the size of a tractor tire, huge and heavy, and required a chain hoist just to lift its sections into place around the propeller. It was positioned in a housing located down a very long, cramped, and dark shaft toward the aft portion of the ship, near that point in the ship where the propeller shaft emerged from the hull. Their work area was so small only one man could fit into the area at any one time. All the while up above, crews continued to practice, unaware that the ship's capability to operate at all was in danger. After nearly two days of effort, Lieutenant Commander Schmidt hit the rack in his stateroom for some sleep.

Machinist's Mate First Class Lee toiled away alone, until he finally got the new bearing into place; inexplicably he made the mistake of failing to inform his boss that he had succeeded. Lieutenant Commander Schmidt continued to sleep in his stateroom. Machinist's Mate First Class Lee knew how tired the officer was when he left to rest, so he thought to leave his boss alone, and allow the man to sleep. It was a nice, personal gesture, but a bad operational decision. The commanding officer of a ship must be informed of the status of his vessel at all times in order to make well-informed decisions regarding the ability of ship and crew to accomplish their mission. When Lieutenant Commander Schmidt finally did wake, he was pleased to learn that Lee had succeeded with the repair. But when he learned that Lee had not yet reported the status of the repair to Captain Seiberlich, but rather allowed the captain to remain ignorant of the status of his ship, Lieutenant Commander Schmidt chewed Machinist's Mate First Class Lee, up one side, and down the other. Machinist's Mate First Class Lee just laughs at that memory today.

Given the evolution of later events in the recovery of Apollo 11, CINCPAC issued personal commendations to Machinist's Mate First Class Lee and Lieutenant Commander Schmidt for their valuable contribution toward the successful operation to recover Apollo 11. If not for Machinist's Mate First Class Lee's skill and dedication to the mission, *Hornet* would have been out of position to conduct the recovery.

Lee downplays the importance of his contribution today. When asked whether family members or friends or neighbors were aware that he played a significant role in the recovery of Apollo 11, Lee scoffed. As far as he was concerned, it was none of their business, and he truly believed that no measure of fuss about his role was warranted. He was just doing his job.

Post-splashdown and recovery, Lt. Clancy Hatleberg and members of Swim TWO await the arrival of USS *Hornet* for the CM retrieval operation. L-R on the flotation collar are Lt. (jg.) Wesley Chesser, SN John M. Wolfram, and QM3 Michael G. Mallory. Hatleberg, in raft #1, has shed his BIG suit and prepares to deep-six the raft and decontamination materials.

Splashdown of Apollo 11. Photo taken by Swim ONE member Mitch Bucklew with a borrowed hand held aerial camera. The film was processed by USS *Hornet*'s photo lab, but never released to the public.

UDT Det Apollo on board *Hornet* after recovery.

USS *Hornet* (CVS-12), configured as the Primary Recovery Ship for the recovery of Apollo 11.

USS *Hornet*'s B&A crane at work during retrieval of the Apollo 11 Command Module, *Columbia*. NASA scientists anxiously awaited their access to her precious cargo of moon rocks, which were stowed on board.

The inflated balloons of *Columbia*'s righting system brought her into a Stable 1 configuration. Note the charred exterior from the fiery re-entry, and the torn kapton foil. Swim TWO commenced the recovery operation just moments after this photo was taken.

Columbia is carefully lowered to elevator #3, where members of *Hornet's* Apollo Command Module Retrieval Team will cut away the flotation collar, then place the spacecraft onto a dolly for transport into the hangar bay.

Members of Swim ONE and Lt. Clancy Hatleberg practice recovery operations prior to splashdown. Helicopter #53 was flown by Cdr. Donald G. Richmond and Lt. William Wesley 'Bill' Strawn of HS-4.

Lt. (jg.) Wesley T. Chesser (L) and SN John M. Wolfram (R) of Swim TWO relax after the recovery of the Apollo 11 astronauts. Lt. Clancy Hatleberg characterized Chesser as "unflappable," and therefore a good choice to lead the team which executed the recovery of Apollo 11. Wolfram was the youngest and the fastest of the UDT swimmers in the detachment.

Lt. Clancy Hatleberg, commanding officer of UDT Det Apollo, flew to Houston in May 1969 to practice egress procedures with the Apollo 11 astronauts. Their BIG suits were designed to prevent contamination of the earth by lunar pathogens.

USS *Hornet* on her final approach to retrieve *Columbia*. *Hornet*'s commanding officer, Capt. Carl J. Seiberlich, personally guided the ship alongside for the retrieval operation, to relieve subordinates of responsibility in the event the module was inadvertently struck and the moon rocks thereby lost at sea.

The Apollo 11 Command Module turned upside down into a Stable 2 configuration shortly after splashdown. The spacecraft's righting system of inflated balloons would require several minutes to right the module into Stable 1 for the recovery operation. Note the charred heat shield from a fiery re-entry into earth's atmosphere.

President Richard M. Nixon shares a light moment with the Apollo 11 astronauts during the welcoming ceremony in USS *Hornet*'s hangar bay about an hour after recovery. *Hornet* + 3 was the motto adopted by the crew of USS *Hornet* to reflect their determination to bring 3 Apollo 11 astronauts safely home.

The astronauts enter a Mobile Quarantine Facility (MQF) within USS *Hornet*'s hangar bay shortly after an HS-4 helicopter transported them from the sea to the ship. Their BIG suits and the MQF were designed to prevent the contamination of the earth by lunar pathogens.

UDT Det Apollo member ADJ3 Mitchell L. "Buck" Bucklew took a photograph of Columbia at the moment of splashdown as HS-4 helicopter #53 hovered above the spacecraft. Bucklew was a member of the cadre class of Navy SEALs. Depicted here during a training session for the recovery of Apollo 11.

The crew of HS-4 helicopter #53 and members of Swim ONE. ADJ3 Mitch Bucklew holds the cover from a drogue chute mortar retrieved after splashdown. Helicopter #53 nearly collided with *Columbia* during its descent through the atmosphere.

FRIDAY, 18 JULY 1969

A full-scale SIMEX was held on Friday, 18 July, just six days before Splash-down day. It began around 0430 and lasted for almost five hours, until 0915. The timing of this SIMEX closely approximated the expected time of splashdown, where HS-4 helicopters would be aloft around 0400 and splashdown was calculated for 0551 in the morning.

The exercise went smoothly, marked by only a couple of glitches. The boilerplate "bounced" off the hull of *Hornet* during the hoisting opera-tion. First, given that boilerplate #1218 weighed approximately 9,000 pounds, a bounce was actually quite a blow. Hornet Module Retrieval Team members who manned training lines were reminded that a blow against either elevator track by a swinging spacecraft could cause sufficient damage to jam the tracks, rendering them useless and the elevator inop-erable. They would have to pay attention to detail and control the beast if it began to swing like a pendulum during the hoist. Second, NASA surgeon Dr. William "Bill" Carpentier was momentarily stunned while jumping from a helicopter into the ocean. Dr. Carpentier's assignment was to care for the astronauts in the event of their illness or injury dur-ing the recovery operation. His participation on Splashdown day would commence from the moment the astronauts emerged from their CM as it floated on the surface of the sea, through the hoisting operation into HS-4's Recovery, and continue as he accompanied the astronauts into the MQF on board *Hornet*. If an astronaut was ill or injured and required treatment immediately after splashdown, Dr. Carpentier was required to deploy with the UDT swimmers helicopter—by jumping.

Carpentier was a competitive swimmer in college and a recreational scuba diver, as well. He was comfortable in the water. And he was expe-rienced at jumping from helicopters into the sea. Carpentier had prac-ticed the skill for several years after his involvement with the Gemini program, and had completed approximately fifty such jumps prior to the recovery of Apollo 11. But this jump proved a challenge. He had not trained with Swim ONE prior to this SIMEX, so he joined their helicop-ter for the practice. Mitch Bucklew, the sea anchor swimmer for Swim ONE, jumped first, from his usual elevation of perhaps ten to fifteen feet above the surface of the water, and at a speed of approximately ten mph. But after Bucklew jumped, a helo crewman directed the pilot to go around for another pass without informing UDT team leader Lt. (jg) John McLachlan that he was going to do so. As the helicopter gained altitude, McLachlan tapped Dr. Carpentier on the shoulder to signal that it was his turn to jump. He did so. But he recalled thinking at the time of the jump that it seemed to be taking an awful long time for him to hit the water. He later calculated that he had jumped from an altitude of about one hundred feet; the impact knocked him almost senseless. Fortunately Mitch Bucklew was close at hand and rendered assistance. But, as Bucklew recalled some thirty-nine years later, Dr. Carpentier swore to "never again" jump from a helicopter into the sea (author interview of Bucklew, 26 November 2008).

Sharks continued to plague the swimmers, but Clancy did his best to downplay the danger. JO2 Chauncey "Chan" Cochran described Hatleberg as a showstopper with his low-key, casual, laid-back style. Nothing seemed to impress or to faze him. Cochran recalled that information related by other officers to reporters during the twice-daily press conferences seemed rather routine after awhile, to the point of being boring. And then Clancy would take the stage. He could deliver lines like, "We got nosed around by a few sharks out there during practice today," with such complete nonchalance that it took reporters a moment to digest what he had said. Hatleberg would continue with an additional four or five comments before some reporter would digest his first remark and shout, "Hey, wait a minute! Did you say sharks?" Said Cochran, "It was the kind of stuff you couldn't stage if you tried. It was pure gold because Hatleberg was everyday heroic in a way that a lot of people at home didn't want to believe about anyone involved in the [military] Service during the Vietnam era" (e-mail from Cochran, 04 February 2008). More sharks were encountered during the operation to recover Apollo 11 than during any other manned spacecraft recovery operation.

Lt. William Wesley "Bill" Strawn served as copilot for Swim ONE, the UDT-12 swimmer team which joined UDT Detachment Apollo just weeks before the recovery of Apollo 11. Strawn recalled that sharks tended to gather and swarm about the floating boilerplate in particularly heavy numbers during night operations, because the spotlight shining down on the water from his helicopter attracted them. From his vantage point high above the action, he saw everything happening down below. His most vivid memories are of UDT frogmen scrambling up the side of the boilerplate and out of reach of the sharks (author interview of William Wesley Strawn, 25 June 2007).

Back home in the United States, Sen. Edward "Teddy" Kennedy was charged with leaving the scene of an accident after his car skidded off the side of a bridge on Chappaquidick Island, near Martha's Vineyard in Massachusetts. The accident resulted in the drowning death of Mary Jo Kopechne, a former campaign staffer for his brother Robert Kennedy. The report temporarily distracted some Americans from their focus upon NASA's moon landing effort. It was Kennedy's other brother, John, after all, who had set the entire manned spaceflight program in motion and provided the direction and inspiration to accomplish the nearly impossible goal of landing a man on the moon before the decade was out—a goal which Neil Armstrong would reach and celebrate in just two days' time, on 20 July 1969. Some believed that John F. Kennedy's untimely and tragic death by assassination on 22 November 1963 served in part to propel his younger brother, Edward, into the political arena as a U.S. Senator representing constituents from their home state of Massachusetts. The awful irony, then, of a tragedy which involved a youthful Sen. Edward Kennedy on the eve of triumph for John F. Kennedy's vision was not lost on the American public.

SATURDAY, 19 JULY 1969

The clock was ticking toward a first landing by humans on the moon, and the return to earth was scheduled to occur four days later. Clancy Hatleberg ensured that his most experienced swimmer team, Swim TWO, exercised in preparation for the big day. SIMEX 4 was conducted, with a simulated arrival of the president. Captain Seiberlich and NASA's team leader on *Hornet*, Dr. Stullken, seemed to agree that full dress rehearsals would best prepare recovery and retrieval teams for the sequence of events which were expected to unfold on Splashdown day.

The crew exercised for five hours, from 0716 to 1214 that day, and employed the NS-50 Tilley crane, mounted on the ship's flight deck to retrieve boilerplate #1218 from the water. They positioned the crane in a starboard forward position for the first retrieval exercise, and port aft for the second exercise. The sea state was a bit rough, so they used the 12-foot nylon strap at the end of the crane's line to ensure enough "play" was available in the line to counteract rolling of the spacecraft into the occasional trough.

UDT swimmers who role-played the Apollo astronauts during the SIMEX experienced a burning sensation in their eyes and breathing problems while inhaling vapors, after their leader, Lt. Clancy Hatleberg, swabbed them down with a sodium hypochlorite solution—a break from standard operation procedure. During BIG training in Houston, Hatleberg was taught to wipe down the astronauts with a Betadine solution. And that is what the astronauts expected from him. But sometime after Hatleberg returned to Coronado, NASA determined that Betadine acted to break down the waterproofing of the BIG suits, a circumstance which completely countered their usefulness as containers to prevent outside persons' exposure to moon pathogens. Hatleberg's BIG procedures were thereafter altered to avoid the use of Betadine in direct contact with the astronauts' BIG suits; sodium hypochlorite was substituted as a decontaminant for the BIG suits. NASA officials tested BIG-suited subjects on board *Hornet* after completion of the SIMEX, and instructed Hatleberg to squeeze excess sodium hypochlorite from his mitt before swabbing down the astronauts. That change minimized the vapors which transited filters in their masks to inflame their eyes and throats.

Following the SIMEX, NASA physician Bill Carpentier, thirty-three years of age, and MQF project engineer John K. Hirasaki, twenty-eight, entered the MQF to begin a pre-recovery isolation period. Carpentier would fly aboard the Recovery on Splashdown day to meet the astronauts after HS-4 hoisted them from their CM for transport back to *Hornet*. Carpentier and Hirasaki would then join the three Apollo 11 astronauts inside the MQF, and remain with them while the trailers were transported back to the Mission Control Center in Houston, Texas. All MQF systems were checked by Carpentier and Hirasaki during the isolation period to ensure they were working.

Hirasaki was a member of NASA's Landing and Recovery Division. In his capacity as project engineer for the MQFs, Hirasaki was responsible for testing and operation of all systems associated with the MQFs.

Given the high visibility of the MQFs in the public eye as a containment facility to safeguard humanity from the spread of possibly lethal lunar pathogens brought back by the Apollo 11 astronauts, assignment as project engineer for the MQFs was a lot of responsibility for a twenty-eight-year-old engineer with less than three years' work experience with NASA. But Hirasaki was up to the challenge. He came from solid stock. John Hirasaki's grandfather was Kichimatsu Kishi, founder of the Kishi Colony located just ten miles east of Beaumont, Texas. A veteran of the Russo-Japanese War which raged at the turn of the twentieth century, Kichimatsu Kishi traveled to the coastal plains of Texas in 1907 to purchase a tract of land on which to cultivate rice and other crops and to found there a colony of Japanese farmers. His was a pioneer spirit. Kishi's son, and John Hirasaki's uncle, Taro Kishi, was the first Asian student to attend Texas A&M University, and a member of that school's championship Southwest Conference football team. John Hirasaki was hired by NASA in 1966. More than forty years later, he continued to work for the agency. He was one of just a handful of NASA engineers who worked in support of both America's first effort to place a man on the moon during the 1960s, and America's second such effort, the Constellation Project, which is ongoing as of this writing. His was an all-American success story, grounded in a belief that formal education was the key to success, when combined with a solid work ethic and an earthy determination.

Moon rocks brought back to earth by the Apollo 11 astronauts were expected to be off-loaded after the retrieval operation and stored in controlled temperature and cryogenic shipping containers for transport back to Houston. NASA team members loaded weighted containers on board one of *Hornet*'s C-1A Greyhound carrier onboard delivery (COD) aircraft and subjected the containers to the stresses of a catapult launch and an arrested landing on the carrier, just to ensure they could handle the shock without cracking. No damage to the containers was observed. Mike Mallory, the strong "workhorse" of Swim TWO from the state of Washington, noted in his diary, "My team had the op today. The whole thing went very smooth—hope the real day goes as good."

Mallory understood there were no guarantees that his team would receive an order to execute the recovery on Splashdown day. NASA's rule concerning deployment of UDT swimmers still held: The first team to reach the splashdown site would ordinarily be ordered to deploy and conduct the recovery. The safety of the astronauts demanded it. But his boss, Clancy Hatleberg, clearly favored the deployment of his most experienced recovery team, Swim TWO, for the recovery of Apollo 11, over any other. He wanted to send his best team to rescue Neil Armstrong, Buzz Aldrin, and Mike Collins, if it was necessary to do so. So Mallory had some reason to hope that he would get his feet wet on Thursday.

The swimmers had "head of the line" privileges in the enlisted mess, which meant, in effect, they did not have to stand in line like everyone else in order to be served a meal, but rather they could walk to the head of the line to "cut in" and eat before others. The privilege was extended to them mostly as a practical measure, to ensure that men who had depleted their physical resources while performing a demanding function which was critical to mission success were quickly fed, to recover and return to service. It was not a professional courtesy extended to them as guests. And it was not an acknowledgment of any kind that UDT frogmen were held in higher esteem than other Navy men.

The swimmers did not abuse the privilege, nor even take full advantage of it. In fact, their officer in charge, Clancy Hatleberg, discouraged them from invoking it at all. But a perception developed and persisted among some members of the crew that UDT swimmers were prima donnas who felt entitled by their superior status to eat first, and to eat while "out of uniform," because they wore wet swim trunks or olive drab shorts with distinctive UDT Blue and Gold reversible T-shirts into the mess hall to dine, while everyone around them wore work uniforms of light blue shirt, and dungaree pants with black Boondocker boots. Bruce Erickson, a member of the #3 catapult crew who was detailed to work in the mess hall during the operation to recover Apollo 11, recalled, "I was aware they had head of the line privileges. I was in line when the UDT swimmers cut in. I think I was fourth in line. What I didn't understand is that a lot of us ;regular' sailors were working long hours also. I never understood why the UDT swimmers were not given access to the officers' or chiefs' mess. Both of those places had much better chow" (author interview of Bruce Erickson, 08 April 2008).

The effort to recover Apollo 11 astronauts from the sea and to retrieve their CM was, in most respects, truly a team effort. But the men who participated in that endeavor could not escape their very human natures. If petty emotion raised its ugly head on occasion to mar what might otherwise have been a perfect accord among them, well, that was to be expected. They were human. But this example of pique was the only such example mentioned or uncovered during months of research, which included the conduct of many, many interviews with men who were there. It is remarkable, not for any measure of pettiness which it may reflect, but for its singularity.

11

Return to Earth

*If I had to single out the piece of equipment that, more
than any other, has allowed us to go from earth-orbit
Mercury flights to Apollo lunar trips in just over seven
years it would be the high-speed computer.*

CHRISTOPHER C. KRAFT, director of flight
operations, Manned Spacecraft Center, Houston

SUNDAY, 20 JULY 1969

MAINFRAME computers used on
earth by NASA to crunch numbers
for navigating to the moon in 1969
were developed by a division of the Sperry-Rand Corporation of St. Paul,
Minnesota, called UNIVAC. UNIVAC replaced the old Germanium
transistors in its mainframe models with something new: silicon transistors. It was the beginning of the silicon revolution. The UNIVAC 642B
was then state of the art. It transferred data at a baud rate of 2,400 bits
per second and had a memory capacity of 5.5 million words. It is simply
amazing to think that men risked their very lives on the reliable operation
of equipment so relatively primitive.

Men first landed on the moon during a time when the very concept
of personal computers was the stuff of science fiction. The manufacture
and mass distribution of pocket calculators to perform even simple mathematical calculations was not yet a reality. Students in high school strug-

gled to master the complexities of a slide rule, a calculating instrument so antiquated that, today it is far more likely to be found in a museum than in an engineer's desk. Telephones were hard-wired at home, in telephone booths, and places of employment; there was no such thing as a cell phone. And the only commonly purchased electronic game was a child's football field which vibrated to make the players move.

The Apollo command module (CM) was a very sophisticated piece of flying machinery in its day. But comparatively speaking, it was but a creaky Model T Ford. The onboard computer was the size of a suitcase with a memory capacity of approximately 70 kilobytes. To put it into prospective, many modern handheld devices today have more than 1 gigabyte of memory, or approximately 100,000,000 times the storage capacity that NASA provided Neil Armstrong and his colleagues to navigate their way to the moon and back. Compared with the Maserati of technological sophistications which we send into earth orbit today, little wonder, then, that Apollo 11 experienced some trouble along the way. Some of it was life-threatening, but viewers and listeners took little note of those reports at the time, though television news anchors like Chet Huntley of NBC and Eric Sevareid and Walter Cronkite of CBS certainly mentioned them as events evolved.

Most often the problems were resolved quickly, leading most members of the general public to assume they were minor in nature—mere glitches and nothing important. The first "glitch" involved a failure of the guidance computer on board Armstrong and Aldrin's lunar lander, *Eagle*, as they descended from lunar orbit to the surface of the moon. Their fellow astronaut, Mike Collins, remained in lunar orbit above, manning *Columbia*, while Armstrong and Aldrin performed the never-before-attempted job of landing a manned vehicle on the surface of the moon. The lander descended at a rate which proved too fast for the guidance computer to handle. The system quickly overloaded with data, and alarms warned astronauts Armstrong and Aldrin that the computer was about to crash. Their own crash, on the surface, could quickly follow.

Fortunately, sitting at a console 240,000 miles beneath them in NASA's Mission Control Center in Houston, Texas, a twenty-six-year-old NASA technician named Steve Bales was on the job. Bales served as the guidance officer, or GUIDO, for Apollo 11's landing on the moon, and he had seen this kind of thing before—during simulations. Bales was supported by a twenty-four-year-old engineer named Jack Garman. Both had only seconds to judge whether the system would indeed overload and crash, leaving the astronauts to deal with the complicated and simply impossible task of manually guiding their own descent to the surface, or whether Armstrong and Aldrin could safely ignore the repeated and persistent alarms which the system generated to warn them of system failure.

Jack Garman had intimate knowledge of *Eagle*'s guidance computer, and he quickly determined that the risk was acceptable. But the final decision to either abort Apollo 11's landing on the moon or to go ahead despite the guidance system's warning rested with Bales, the GUIDO. Garman's judgment regarding the alarm was "that's okay"; Bales told flight director Gene Kranz to go ahead with the landing, and the entire issue was resolved in a matter of seconds.

Months later, after excitement surrounding NASA's successful effort to land men on the moon subsided, the president of the United States hosted an award ceremony to recognize NASA's accomplishment, during an eight-year journey, to develop the manned spaceflight program which reached John F. Kennedy's goal of landing a man on the moon before the decade was out. The award was a NASA Group Achievement Award, presented to recognize the efforts of the entire mission operations team. But one man was selected to represent all of NASA at the ceremony and to accept the award from President Nixon. That man was not Neil Armstrong, the first man to actually walk on the moon. It was not Dr. Thomas O. Paine, the director of NASA. And it was not Gene Kranz, the Apollo 11 flight director at NASA's Mission Control Center in Houston. It was twenty-six-year-old Steve Bales. President Nixon said of Bales, during the award ceremony, "This is the young man, when the computers seemed to be confused and when he could have said Stop, or when he could have said Wait, said, Go."

As they approached the surface of the moon, the astronauts sought a safe landing site. It was not enough to simply plunk down into a crater and shut off the engine. *Eagle* required a relatively flat and stable surface, not only to remain upright after landing, but also to ensure proper liftoff after Armstrong and Aldrin completed their moon walk and would blast into lunar orbit to join with Mike Collins in *Columbia*, and return to earth. If *Eagle* canted too severely after landing, leaning over too much in any direction, then liftoff itself might have been either entirely impossible to achieve, or might have required more fuel than *Eagle* carried on board. Armstrong and Aldrin could find themselves stranded on the surface of the moon, with no hope of rescue. But a smooth, flat landing site was not easily found. NASA selected in advance an area which appeared to offer plentiful sites, but the view from earth apparently lacked the clarity and detail of Armstrong's view through his window inside *Eagle*. From his vantage point, the astronaut saw huge rocks the size of houses and deep craters. Landing among them was not an option.

At the last moment before *Eagle*'s guidance system directed her to touch down on the surface, Armstrong assumed manual control of the spacecraft and flew her beyond danger. A good decision on his part, but the astronaut soon faced another challenge: low fuel. There are no gas stations on the moon, and *Eagle* carried a limited supply on board. Armstrong was able to maneuver the craft beyond the field of rocks and craters below, but from his perspective inside *Eagle*, the field seemed endless. He flew on. And down below, in the Mission Control Center in Houston, engineers measured the remaining fuel in *Eagle*'s tanks, not in gallons or liters, but rather in minutes and seconds of operating time. Like Mario Andretti, who won the Trenton 200 motor race in New Jersey just the day before, Neil Armstrong was running out of gas. Andretti finished his race with just one gallon of fuel remaining in his tank. Armstrong had but eight seconds of flight time remaining when he spotted a clearing and quickly set his lunar lander, *Eagle*, down on the surface of the moon. Shortly thereafter, he informed Houston that "*Eagle* has landed." Armstrong's transmission sparked a jubilant roar through Mission Control Center in Houston; the center was briefly transformed from a tightly controlled atmosphere of intensely focused geeks to a jubilant celebration of excited young kids.

NASA had succeeded in placing men on the moon. Soon after Armstrong scaled the steps of the lander and hopped onto the surface. The dust from his first step had not settled when he transmitted his immortal words: "That is one small step for a man; one giant leap for mankind." Armstrong was soon followed by Aldrin, and the two men remained on the surface for nearly two hours gathering rocks and fulfilling their mission. As they prepared to depart another potentially life-threatening glitch occurred.

Eagle had but one main engine for liftoff. And no one knew for a certainty that it was actually going to fire when directed to do so. No one had ever landed on the moon before, and consequently no one had ever actually tested the system in that environment. Mission Control was just crossing its fingers and hoping for the best. It would have been the height of irony, and the worst of tragedies for Armstrong and Aldrin and their families, had that candle failed to light. It was one thing to celebrate the accomplishment of the first half of President Kennedy's stated goal—"before this decade is out, of landing a man on the Moon"—it would have been quite another to mourn the loss after failing to accomplish the second half of President Kennedy's national goal—"and returning him safely to Earth." But that is nearly what happened.

The electrical wiring for *Eagle*'s main liftoff engine was purposefully designed to prevent accidental and premature ignitions of the engine and to prevent other equally disastrous mishaps from occurring. The simple expedient for such preventive design was to build a gap into the wiring. An astronaut might accidentally hit the ignition button, but the electric current would end at the gap, before reaching the engine itself. The upshot: no accidental ignition of the engine, no firing of the rocket, and no accidental liftoff before astronauts were seated and ready to leave the surface of the moon.

To close that gap in the electrical wiring, though, when astronauts were in fact ready to leave the surface of the moon, NASA engineers incorporated a button into a circuit panel within the astronauts' cabin. Armstrong and Aldrin had only to depress that button so that it was flush with the circuit panel to close the gap, which would allow an electrical current to flow between the ignition button and *Eagle*'s main liftoff engine. Then they could leave the moon. The button was, however, broken. Armstrong and Aldrin wore bulky spacesuits during their walk on the surface of the moon. When the two astronauts stepped back into the lunar lander after their two-hour moon walk, and in preparation for their departure from the surface of the moon, one of them turned around suddenly within the cabin and a portion of his bulky spacesuit snapped that button off the circuit panel. Armstrong and Aldrin were potentially stuck on the moon, with no means of firing their liftoff engine.

Neither Armstrong nor Aldrin panicked. Buzz Aldrin produced his specially designed, NASA-approved astronaut pen, and simply used it as a replacement tool to push a plunger into place and complete the electronic circuit. He did not need a button; he had a brain. NASA had taken guff from government bean counters and others when they first proposed to develop such a pen for use by their astronauts. The need for a pen that would operate in a zero-gravity environment while held in the possibly gloved hand of an astronaut seemed clear to NASA. But bean counters viewed expenditure of limited research and development funds on something as mundane as a writing instrument—when so many pens and pencils were already available on the open market—as a waste of resources. Fortunately NASA prevailed. That pen saved the lives of Neil Armstrong and Buzz Aldrin by enabling them to lift off from the moon and return safely to earth. It was a good investment.

On earth, the reaction to NASA's successful moon shot was predictable. People had gone to great lengths to observe firsthand as Neil Armstrong and Buzz Aldrin romped around the surface of the moon. Some twenty South African businessmen chartered a commercial aircraft to fly them to London, just to watch the moon walk on TV; South Africans did not yet have television in their country. A Tokyo beer garden installed five high-powered telescopes and hired an astronomer to help patrons peer into the heavens toward the moon throughout Apollo 11's flight—the charge: buy a beer. Pan American Airways reported that 16,700 customers purchased reservations for flights to the moon—the company predicted that commercial travel to the moon would be available by the end of the century. In Farragut State Park in Idaho, 30,000 Boy Scouts attending the seventh National Boy Scout Jamboree cheered when they learned that Neil Armstrong and Buzz Aldrin, both of whom had been Boy Scouts, landed safely on the moon. The jamboree had opened on 16 July, the day that Apollo 11 launched into space, and the boy scouts followed radio reports of the mission's progress from that day.

Not everyone witnessed the landing. Nearly one quarter of the earth's population was purposefully kept in the dark about the successful voyage to the moon. The governments of North Korea, North Vietnam, Albania, and the People's Republic of China deliberately blocked media coverage of the event, thus denying that knowledge to approximately 800 million people. Ironically, the People's Republic of China developed its own manned spaceflight program much later, and was projected to become the next nation to land men on the moon.

And some people were not happy about television coverage of the flight of Apollo 11 at all. In Ottawa, Canada, CJOH-TV reported that fifteen viewers called to complain that the station preempted their favorite TV show on Sunday night to broadcast footage of Armstrong and Aldrin on the moon. Their favorite TV show? *Star Trek*.

On *Hornet*, steaming in the Pacific en route to the primary splashdown zone, the ship held to a holiday routine. It was a Sunday, and watch lists were cut to a bare minimum to allow free time for as many men on board ship as possible. Recovery and retrieval exercises were not held. The officer of the deck (OOD), who steered the ship from the bridge, maintained a logbook of significant events: the deck log. Young *Hornet* officers took turns manning the bridge as OOD and made entries into the ship's deck log throughout their watches. Squeezed in between two routine entries in *Hornet*'s deck log for Sunday, 20 July 1969, was an entry by Lt. (jg) James S. "Steve" Lauck, OOD, which documented Apollo 11's landing on the moon. Lauck's entry for the 0800–1200 watch on the bridge that day was as follows: "08-12 Underway as before. 0918 Apollo eleven lunar module has landed safely on the moon. 0935 Received daily draft report: FWD, 28ft 2 in, AFT 28ft 2in, MEAN 28ft 2in, Gross tonnage 38,147." It was a typically Navy, just-the-facts-ma'am, official entry into a ship's deck log. Lauck recalled much later that the rest of the ship might have been wildly celebrating the event, but for men on the bridge of the ship, it was all business. No high fives or other emotional displays. He was running the ship, and therefore shouldered a heavy responsibility for the safety of ship and crew throughout his watch.

On the flight deck Lt. Richard J. "Rich" Barrett, the pilot of HS-4 helicopter #64, designated by call sign Swim TWO for the recovery operation, took advantage of the break in training to go jogging. Barrett normally ran about twelve miles around the flight deck—quite a distance. Pilots of Navy helicopters during the Vietnam War conducted rescue operations to recover downed pilots up to twelve miles inland from the coast of Vietnam. Barrett's rationale for jogging a standard twelve-mile distance was quite simple: In the event his helicopter ever went down behind enemy lines twelve miles from the coast, and he found himself pursued on foot by North Vietnamese forces, Barrett reasoned he would simply "outrun the bad guys to the coast." Barrett was usually joined during his jog by UDT frogman Lt. (jg) Wesley Chesser, the leader of Swim TWO. The UDT man was in exceptional physical condition, and he demonstrated it while jogging. Said Barrett, "[Chesser] would be running *backwards*, expecting me to hold a conversation with him. I'd try, but I always hated it when he did that! I'd have on running shoes and he'd be running in boots! It was a little embarrassing."

Elsewhere on board the ship, celebrations of the moon walk did occur. RM2 Roger W. Beahm, a radioman from Wisconsin, recalled that *Hornet* provided continuous radio updates for the crew, so everyone knew that the landing was about to occur. There was no TV coverage for the crew of the *Hornet*, which operated at sea at the time, but there was great interest in the landing nevertheless. An announcement about the moon landing was made over the 1MC public address system, and a big cheer erupted throughout the ship. But very quickly, Beahm recalled, exhilaration was replaced by a cold reality: "We knew that we were on deck, big time. Things got serious after that, and intense." The crew realized that the world was about to turn its attention to their efforts to recover Armstrong, Aldrin, and Collins from the warm swells of the Pacific Ocean, and, with so many people watching their actions, they simply had to succeed. Lt. (jg) Richard F. "Dick" Powers III echoed Beahm's recollection: "From then on, for the crew it was countdown to kickoff. Time then passed very quickly. There was little sleep, and now we had a laser focus on our mission."

When asked to recall where he was, exactly, when he first learned that Neil Armstrong had stepped foot on the moon, MM3 Clifford R. Burr of Texas responded without hesitation. He remembered exactly where he was. Burr was an enlisted man, a machinist mate who specialized in hydraulics, and who served with the Hornet Module Retrieval Team to correct hydraulics-related problems with the NASA winch as it hoisted the boilerplate or CM aboard ship during its retrieval from the sea after splashdown. Burr was standing watch alone, in an elevator pump room far belowdecks, reading a book titled, "The Happy Hooker," when the announcement was made over the 1MC.

That night many of the men on board *Hornet* gravitated to the flight deck to take in the night air and contemplate the days ahead. The moon shone brightly that night and appeared larger and closer to the eye than ever before in the refracted atmosphere of the Pacific Ocean. Lt. Cdr. William G. Tasker, the ship's meteorologist, walked the deck with a NASA technician. He recalled it was a clear night, and as they looked up at the moon, the NASA technician observed with a sense of scarce belief and wondrous awe in his voice that "our guys are up there." EM3 Michael Lee Laurence, one of two enlisted electricians responsible for lighting on the flight deck, had no need to apply his skills that night. He, too, contemplated the near future: "Wow, those guys are up there now, but they're going to come back and land near us."

Not everyone celebrated solely by contemplating the moon that night. JO2 Chauncey "Chan" Cochran, the enlisted journalist from Ohio who served to support the press pool on board *Hornet* throughout the recovery operation, was seated at his desk in the media center that night, filing stories for reporters when the phone rang. One of the chaplains on board was on the line and ordered Cochran to report to a stateroom assigned to reporters. Upon Cochran's arrival, someone handed him a twelve-ounce glass of Jack Daniels. A party was in session and Cochran was invited. Later on Cochran went out to the flight deck and lay down to look at the moon. He imagined the astronauts up there, streaking even then toward their rendezvous with *Hornet* in the middle of nowhere, the Pacific Ocean. Cochran recalled the scene was surreal at night in the middle of the ocean, surrounded by nothing but darkness, with the moon so full and bright above him.

Two great vessels, then, were charting separate paths toward a common rendezvous point in the Pacific Ocean. From above, *Columbia* and her service module prepared to exploit a slingshot effect to tear away from the gravitational pull of the moon and travel in excess of 25,000 mph toward the splashdown point on earth. Below, *Hornet* neared her initial splashdown point, where she would take up station to assist the astronauts of Apollo 11 after splashdown. To achieve complete success, the two vessels would have to meet in the same spot of ocean, and at the same time.

The crew of Apollo 11 had experienced their share of technical glitches and problems along the way, but the worst was behind them. Aided by an onboard inertial measurement unit (IMU), the crew augmented position readings provided by NASA's earth-based worldwide radar system with their own onboard star sightings obtained by sextant, to calculate an exact fix on their position in the universe.

The IMU was developed under contract for NASA and General Motors' AC Spark Plug Division by the Instrumentation Laboratory of the Massachusetts Institute of Technology (MIT)—the same academic institution that developed the guidance and navigation system for the Navy's Polaris missile. The IMU was spherical in shape and larger than a basketball. It housed a combination of accelerometers (to measure speed) and gyroscopes (to measure direction of travel), plus temperature gauges. It worked by sensing its own rate and direction of motion and then providing that data to an onboard computer, developed under contract by the Raytheon Company, to track its position by means of dead reckoning. Apollo's onboard computer was the first to use integrated circuit technology, the same technology employed later to develop desktop computers. Similar tracking data were provided by NASA's worldwide radar system and automatically transmitted to Apollo 11's computer. And the entire system was augmented by the astronauts themselves, who took star sightings through their onboard sextant and then submitted the data to their onboard computer.

Apollo 11's onboard computer was "hardwired" in the truest sense of the word. Every navigational equation required to track Apollo 11 to and from the moon was hardwired into the system by hand—a true hard drive. Variable input (readings) from the IMU, from NASA, and from the astronauts were plugged into the equations, calculated, and the results were compared against NASA's pre-programmed flight path to and from the moon. If calculations indicated that the Apollo 11 spacecraft was slightly off the pre-programmed flight path, then signals were automatically sent to the craft's maneuvering system to adjust the craft's relative motion and bring it back on course.

Apollo 11's navigation system worked as designed, and she remained on course throughout the remainder of her trip. *Hornet*, however, experienced some difficulty in meeting her end of the bargain as the two vessels maneuvered toward their common rendezvous point in the Pacific Ocean. *Hornet*'s issues had nothing whatsoever to do with Machinist's Mate First Class Lee's troublesome bearing, however. The problem was navigation.

MONDAY, 21 JULY 1969

Sextants have been used since before the days of Christopher Columbus to navigate the oceans. They are sophisticated instruments, easily mastered with practice, and so reliable and accurate in producing input for navigational calculations that men required nothing else to fix their positions on the globe for hundreds of years. In 1969 the vast majority of ships then plying the oceans relied almost exclusively on the sextant to figure out where on earth they were.

The first GPS system, a Navy satellite system, was activated, and onboard systems were deployed to the fleet just before USS *Hornet* departed her home port of Long Beach, California, to recover Apollo 11 from her splashdown in the Pacific Ocean. Perhaps three U.S. Navy vessels were equipped with the system at that time. *Hornet* was not one of them. She continued to rely on the sextant and on fathometer readings, which located seamounts on the ocean floor, to figure out her position on the globe.

Hornet arrived at her initial splashdown point, about 1,000 miles southwest of Hawaii, on Monday, 21 July 1969, three days before the flight of Apollo 11 was scheduled to end in a splashdown on the sea nearby. The rendezvous point with Apollo 11 was in a very remote, uncharted area of the Pacific Ocean. *Hornet* had not seen another ship or plane for the previous six days while en route to that spot. Clouds covered the sky and blocked her view of the heavens. Star sightings by sextant were impossible, so she navigated for the next three days by dead reckoning—a manageable challenge for skilled and experienced sailors.

Hornet had on board a solid team of experienced navigators, including her commanding officer, Captain Seiberlich, but also her XO, Cdr. Chris Lamb; the assistant navigator, Lt. Cdr. Edward M. "Ned" Dunham; and a division of enlisted quartermasters led by QM1 Howard B. Mooney. The ship's new navigator, Cdr. Robert A. Costigan, had little at-sea experience as a navigator, but he had at his disposal a stable of qualified assistants. Rumor had it that his most able assistant, Master Chief Quartermaster Cave, had less than two weeks earlier departed the ship while it was in port in Pearl Harbor after experiencing a disagreement with his boss.

During the previous several days, as *Hornet* transited to her initial splashdown point, she scanned the ocean floor in the relatively uncharted area of the Pacific Ocean, employing fathometers which reached to an average depth of 18,000 feet. The mapping effort discovered a number of new seamounts, underwater mountains which rose thousands of feet above the sea floor. Those mountains, when fixed on navigational charts, could be used to assist navigators as they charted their way across the sea. Assistant navigator, Lt. Cdr. Ned Dunham related, "We've found six new mountains. The highest peak rises 11,000 feet from the floor of the sea. We have the latest charts, and they don't have any of the ones we found."

Naval custom entitled ships that discovered new seamounts to name them. One newly discovered seamount was located just five miles west of the designated splashdown point for Apollo 11, and Dunham thought it appropriate to name it Apollo 11. Another was tentatively named *Hornet*'s Hump. And Dunham thought that three others would be most appropriately named after each of the Apollo 11 astronauts. "We will chart the area thoroughly in the next few days, and if we should lose the stars in the clouds, it will give us an excellent navigational fix," Dunham said (USS *Hornet* message, Presrel—Daily Wrapup, 21 July 1969 [courtesy of Cpt. Chris W. Lamb, USN (ret.)]).

Above them, the crew of Apollo 11 remained in lunar orbit but prepared for their return to earth. They were joined in orbit by the Soviet unmanned probe, Luna 15. The Soviet's remotely piloted robot was launched several days before the launch of Apollo 11, but its speed was slower than that of Apollo 11 and it followed a flight path which required a greater distance to travel to the moon. It therefore entered lunar orbit hours after Armstrong and Aldrin had walked on the moon and returned to join Mike Collins in the Apollo 11 command module, *Columbia*, in lunar orbit. On earth, American news networks fretted quite openly that the Soviet craft might yet scoop up samples of lunar soil and return them to earth for examination before the Apollo 11 splashdown, thereby "scooping" Apollo 11 as well and stealing American thunder in the process. But something went wrong with the craft as it descended toward the lunar surface. It crashed into the surface at a speed of three hundred mph, approximately one hundred miles from the remains of the abandoned lunar module, and was destroyed.

Back on earth, *Hornet*'s SIMEX went well. The only problems encountered concerned overheating by the swimmers and discrepancies in the SARAH bearings reported by all airborne units. The latter was not a minor matter. Sea water temperatures in the Pacific Ocean ranged from 83° to 86° Fahrenheit—almost as warm as bathwater. UDT frogmen involved in the recovery operation wore their standard-issue wetsuits to protect themselves from caustic fluids which flowed down the sides of Apollo command modules after splashdown. The combination of warm water, warm air temperatures, thick wetsuits that insulated their bodies, and strenuous exertion during an unusually long SIMEX of four and a half hours quickly overheated and tired the swimmers. But there was no ready solution to the problem. And, typically, they did not complain.

The SARAH bearing discrepancies were more troublesome. Radar equipment employed by *Hornet* had been thoroughly overhauled and recalibrated by NASA and *Hornet* technicians before the ship departed her home port of Long Beach, California, on 26 June. Accuracy of radar readings was obviously a priority for NASA. But bearings produced by *Hornet* during the SIMEX to the targeted boilerplate differed from SARAH bearings developed by all airborne assets employed during the SIMEX. During an actual recovery, which was in essence a tactical situation, HS-4 helicopters would almost certainly rely on their own SARAH-produced bearings to find the CM before she splashed down into the sea, rather than depend on *Hornet* to vector them to the target. But the discrepancy would have to be resolved.

Neil Armstrong received a letter on board *Hornet* that day, though he did not know it yet. Tom E. Slater of Chanute, Kansas, addressed his envelope to Commander Armstrong, Apollo 11, Splashdown, USA. Mail carriers must have figured out fairly quickly that Splashdown USA was for all practical purposes USS *Hornet*, because it took only five days for the letter to reach the middle of the Pacific Ocean from Kansas. NASA officials placed the letter inside the MQF for Armstrong to read on the voyage from recovery back to Hawaii.

That afternoon a COD flight brought to *Hornet* from Johnston Island a special delivery for young Lt. (jg) Dick Powers III. Powers was designated to serve as USS *Hornet*'s escort and aide to President Nixon while the commander in chief was on board for the welcoming ceremony for the astronauts. That was quite an assignment for a twenty-three-year-old man just barely out of college. Among his duties was to safeguard three very special ball caps, which the president intended to present to the heroic astronauts of Apollo 11 during the welcoming ceremony in the hangar bay after their splashdown and recovery. The cap brims were emblazoned with "USS *Hornet*," the ship's motto for the recovery, "*Hornet + 3*," and their names.

Powers was naturally a bit nervous about his role, but he was also excited. He recalled, for example, that he "must have checked (at least) 100 times" with the XO to ensure he understood correctly that he was to meet with the president in the flag plot on Splashdown day and not somewhere else in the superstructure of the ship. His order on receipt of the Presidential ball caps was to secure them in a safe until he met with the president on Splashdown day. But Powers was still just a kid. He was so excited when first viewing the ball caps that his first thought was to share the excitement with roommates Lieutenant (jg) Lauck, who made the entry into the ship's deck log when the astronauts landed on the moon, and ship's public affairs officer, Lt. (jg) Milton "Tim" Wilson III, the heavyweight wrestler from Stanford University.

Powers called his roommates and asked them to meet him in their stateroom. The three young men donned flight suits and mugged for one another with Wilson's camera as they played with the presidential gifts. It was a typically sophomoric prank, to be expected perhaps of junior officers. But if anyone else had learned of their actions, heads would have rolled. They still have the photos today.

In a less jovial spot on earth at that time, North Korean military infiltrators attacked an American Army 2nd Division outpost across the demilitarized zone (DMZ), which divided North and South Korea, with small arms fire and grenades. There were no casualties. North Korea was among a handful of nations on earth that prevented their people from learning about Apollo 11's successful moon landing. The government of North Korea was apparently not impressed with either American military or space prowess. Peace and security were never a "given." Americans remained vigilant for signs of danger all around them.

Hornet occasionally sent aircraft aloft to check weather conditions within her immediate operating area, but she had no means to collect weather data beyond the limited range of such flights. Conditions which developed far over the horizon were hidden from view. And commercial weather satellite systems simply did not exist in 1969. So it was impossible for *Hornet* to predict weather conditions days in advance of the splashdown.

Fortunately a young U.S. Air Force captain, Hank Brandli, assigned to a secret intelligence unit on Hickam Air Force Base, in Hawaii, had the ability to do so. What he saw alarmed him. A trained meteorologist from MIT, Brandli's mission in 1969 was to support a U.S. intelligence community program to exploit photographs taken by one of the nation's first spy-in-the-sky electro-optical weather satellites. The very existence of the satellite was known to only a handful of people at the time, and its employment by the intelligence community and its capabilities were extremely sensitive. Intelligence "product" in the form of images taken by the satellite were classified Top Secret, and were handled within the intelligence community only by those who were specially indoctrinated into the program. The satellite was capable of taking refined photos of specific target areas on earth—aerial photographs over naval bases in Russia, for example, or air force bases in the People's Republic of China and similar targets in Vietnam. If a weather system rolled into a target area and covered it with clouds, intelligence analysts found themselves staring in frustration at layers of clouds rather than their intended targets.

Brandli's function was to study weather patterns in order to predict when specific target areas might be clear enough to photograph. The National Reconnaissance Office (NRO) then used Brandli's predictions to guide American spy satellites over intelligence targets. In a sense Brandli was among the first meteorologists to employ satellite photographs to predict the weather. Though he did not work in support of the Apollo 11 recovery effort, Brandli was aware that Neil Armstrong and Buzz Aldrin set foot on the surface of the moon on 20 July and that Armstrong, Aldrin, and Mike Collins were scheduled to return to earth and splash down in the Pacific Ocean on 24 July. As he studied satellite photos of the splashdown area, Brandli realized that a storm front which included particularly dangerous thunderclouds and vortex-like winds was moving toward the splashdown area, and he predicted that the front would reach the splashdown point just as the Apollo 11 astronauts were scheduled to arrive.

Captain Brandli had seen this type of system before. He referred to the towering thunderclouds as screaming eagles, a moniker which reflected their general shape as he viewed them from above and also their ferocity. The clouds climbed to an altitude of 50,000 feet. Brandli understood that *Columbia*'s three main parachutes would be torn to shreds by the howling winds of the storm as they descended through the clouds. Without its parachutes, the CM would drop like a rock into the ocean. The astronauts would never survive the impact.

The photographs which Brandli reviewed were classified Top Secret. He could not share them with just anyone. But an alert to warn NASA and the Apollo 11 astronauts of the danger which they faced had to be issued, regardless of the sensitivity of the intelligence community's secret satellite program. Brandli chose to bypass the bureaucratic chain of command and contact someone who could take immediate action to save the astronauts: Navy captain Willard S. "Sam" Houston Jr.

Houston served as the commanding officer (CO) for the Fleet Weather Center at the Naval Base, Pearl Harbor. As a collateral duty, Captain Houston additionally served as the senior meteorologist for R. Adm. Donald C. "Red Dog" Davis, the commander of Manned Spaceflight Recovery Forces, Pacific. In the latter capacity, Captain Houston served, in essence, as the weatherman for Apollo recovery forces then at sea in the Pacific Ocean.

By a happy coincidence, Captain Houston served during a prior assignment with the Office of the Joint Chiefs of Staff in the Pentagon and as a member of the board which developed the special spy satellite program. He was cleared for access to Captain Brandli's information. The two men met in a parking lot at Hickam AFB. Once alerted by Brandli, Captain Houston's responsibility was to alert Rear Admiral Davis to the danger posed by screaming eagles thunderclouds to the astronauts of Apollo 11. Davis had no choice other than to forward the warning up his chain of command, with a recommendation to change the splashdown site to another, perhaps safer region of the ocean. The Navy could not make the decision to alter NASA's plans. Only NASA could do so.

The information that formed the basis for the admiral's recommendation was derived from a source so sensitive that its very existence could not be revealed to decision makers in Washington, D.C., without violating national security: this only complicated the issue. Someone within the U.S. intelligence community would have to authorize the disclosure of sensitive source information to senior NASA officials who were not yet privy to the secret spy satellite program. NASA officials would then have to decide whether to stick with the original plan or to alter the splashdown target zone. The astronauts were just preparing to leave the moon and begin their flight back to earth. There was little time to accomplish much.

Rear Admiral Davis cautioned Captain Houston to brace himself for a possible backlash in the event his interpretation of the weather satellite photos was in error. NASA would have to recalculate the CM's re-entry to a new splashdown site; the Navy would have to direct its primary recovery ship and other recovery forces to steam to that new site; and the president of the United States, who expected to be on board USS *Hornet* to welcome the returning astronauts home, would have to alter his travel plans. If the alarm proved false, it could cause irreparable harm to the careers of both men. Rear Admiral Davis remarked, "You'd better be right, young man," and issued orders to alert NASA to the danger.

TUESDAY, 22 JULY 1969

At 1157 EDT, the astronauts of Apollo 11 fired their engines to blast out of lunar orbit and began their trip home. Far below in the Pacific, *Hornet* conducted a full-scale SIMEX that day, including a simulated arrival of the president. The exercise commenced at 0419 and concluded some four and a half hours later at 0847, and employed the B&A crane for the retrieval operation. That was the crew's final and eighteenth exercise conducted before Splashdown day. Before the exercise commenced, however, Captain Seiberlich exhorted his crew to give their best effort yet. He made reference to the extraordinary astronauts of Apollo 11. "Those men," he said referring Armstrong and his crew, "who will be coming down to us are professionals. Let's show them that in our sphere, which is the sea, we are every bit as professional" (author interview of Peter D. Beaulieu and review of Beaulieu's notes).

They were well prepared and ready. Captain Seiberlich had remained true to his word. The crew would retrieve that thing as they practiced, and he had drilled them to a fare-thee-well. "We're trained up to the correct point," Seiberlich declared. "We're not overtrained. We still have our best effort in us, and we'll put that out Thursday morning [Splashdown day]." When asked to describe his memories of the efforts made by the Hornet Module Retrieval Team to prepare for the recovery of Apollo 11, their team leader, weapons officer, Cdr. Harley L. Stuntz, simply said, "I just remember that it was a lot of work. We worked almost 24 hours per day. Little sleep. I was tired all the time" (author interview of Stuntz, 19 July 2007). It was time to stop practicing, and to start doing.

During the exercise that day, the recovery helicopter, which transported swimmers role-playing the astronauts of Apollo 11 during the exercise, touched down on the deck of *Hornet* at exactly the same moment that Neil Armstrong triggered his rockets and began his return trip to earth. "Being a seafaring man, maybe I'm superstitious," said Captain Seiberlich, "but to me that's a good omen."

Hornet was joined in the Pacific by USS *Hassayampa* (AO-145). The two ships commenced underway replenishment operations at sea (UNREP), because *Hornet* needed both fuel and other supplies to meet mission requirements. Providing those necessities was *Hassayampa's* function within the fleet. This was a routine operation, but nevertheless a risky and somewhat dangerous one. It requires two ships to operate at ten to twelve knots along parallel courses while lines are established between them for the transfer of fuel oil and dry goods. The distance between ships during such operations is typically no more than 100–125 feet, which is practically spitting distance for sailors at sea. There is little room for error, and it takes calm nerves and a steady hand at the wheel to avoid collision. The wake and wash created by each ship as it plows ahead through the water enters the zone between them to complicate maneuvers and can act to draw them toward one another. The fact that most aircraft carrier UNREPs were conducted at night only adds to the difficulty of accomplishing such operations safely. During her most recent WestPac cruise, *Hornet* had collided with another ship during an underway replenishment operation and caused superficial damage to the other ship's superstructure. She could not afford a similar accident this date, with the recovery of Apollo 11 scheduled to occur in less than forty-eight hours. Besides, a flag officer was about to join them. And nothing could ruin a CO's day more quickly than the rueful eye of an admiral on deck. The underway replenishment was accomplished without incident, and *Hornet* took on more than 800,000 gallons of fuel. She would need every drop of it, and soon. Rear Adm. Red Dog Davis, CTF-130, was welcomed aboard.

Rear Admiral Davis arrived before the other VIP guests, most likely to receive a status report from Captain Seiberlich and the NASA recovery team on board *Hornet* concerning their preparations for the recovery. Davis was operationally in charge of military forces involved in NASA recovery operations in the Pacific theater, and he was therefore both responsible and accountable for their efforts. Success would shine on him, and failure would fall on him with equal measure. "Normally, I wouldn't be here," he told the press corps following his meeting. "I'd be at my headquarters in Kunia [Hawaii]. I chose to be here because this is such a fantastic event. There has never been an occasion when the recovery ship was as ready. This crew is ready for all possibilities," he concluded in an effort to assure them (author interview of JO2 Chauncey "Chan" Cochran, 6 February 2007).

The ship had no flag officer quarters reserved for Rear Admiral Davis, so Commander Lamb gave up his stateroom for the duration of Davis's stay. Davis would remain on board throughout the recovery and retrieval operations. Lamb found space with the CO's executive assistant, Lt. Henry Francis Dronzek, and several other junior officers in a very cramped junior officer stateroom until after the recovery.

The Hornet Module Retrieval Team put a new winch line on the B&A crane in anticipation of the Big day. They tested it by lowering the boilerplate into the water, then hoisted it back onto the deck. There remained plenty of work to do, but the pace and intensity of preparations waned as Splashdown day neared.

The boilerplate was taken below to the hangar bay and connected to a transfer tunnel that was erected at the prime mobile quarantine facility (MQF), which was expected to be used by the Apollo 11 astronauts after they boarded the ship. A transfer tunnel, which was essentially a plastic see-through tunnel, extended for a distance of perhaps ten feet between the primary MQF and the retrieved CM and would be used during transfer of moon rocks from the CM to the MQF to prevent moon germs from escaping into the hangar bay and infecting members of the crew and civilians on board the ship. The tunnel was not completely sealed, and it appeared to some that the tunnel served a greater cosmetic and public relations function than a true laboratory function to absolutely prevent infection by containing the spread of moon germs. It looked like a gimmick to minimally satisfy the Interagency Committee on Back Contamination's (ICBC) procedural requirements to prevent the spread of moon plague to the populace. But whatever the functionality of the transfer tunnel, the placement of boilerplate #1218 in the CM's final resting position in the hangar bay on 22 July 1969 was simply a test to identify kinks or glitches that might arise during such a procedure, so that solutions could be developed before Splashdown day. NASA officials found that turbulence and wind through the hangar bay could pose a problem, but steps were taken to prevent damage to the tunnel in the event bad weather conditions on Splashdown day proved to be a problem. Dr. Carpentier and MQF engineer John Hirasaki reported that they were feeling fine after their stay inside the MQF and that all systems appeared to be functioning nominally. One item on their checklist was a test of communications equipment to be used during the welcoming ceremony with President Nixon.

Inside the ship, a talent show was broadcast to the crew via its internal closed-circuit TV3. The sales division, headed by Lt. (jg) Tyler J. Bateman, sponsored an advertisement for a special Splashdown day sale on novelty items—specially ordered Apollo 11 pens, keychains, beer mugs, and baseball caps. Sales exceeded expectations, and many of those items can be found today, boxed and stored by *Hornet*, HS-4, and UDT veterans in garages and attics throughout the nation.

A discussion ensued between John C. Stonesifer, Dr. Stullken's assistant on the NASA recovery team on board ship, and Lt. Clancy Hatleberg, the BIG swimmer who was scheduled to decontaminate the Apollo 11 astronauts with sodium hypochlorite solution after splashdown. Throughout the nearly eight-year history of NASA's manned spacecraft program, that procedure had never been employed because, of course, no one had ever set foot on the moon before. They laughed at the thought of TV cameras zooming in for worldwide coverage of Clancy and the astronauts, sitting in their little raft as it floated on the open ocean swabbing one another down with mittens. Stonesifer had to chuckle. "It will be quite a scene," he predicted.

WEDNESDAY, 23 JULY 1969

It was not a good way to start the day. At 0431 a small fire broke out in compartment B-306-L, just one level below the flight deck, actually within the struts. Insulation on a cable used to start aircraft engines began to burn. Crewmen had it out within minutes and the scene was secured by 0444. The *Fort Wayne Journal-Gazette*, 22 July 1969, reported the incident and readers may have concluded that the ship sent to recover the Apollo 11 heroes was burning, sinking, or falling apart at sea, the day before they were supposed to rescue the astronauts. But Commander Stuntz, weapons officer, declared that it was a sensational report concerning a routine matter. Fires aboard ship can be disastrous and are always taken seriously. But that one posed no threat to the ship or crew.

USS *Arlington* (AGMR-2) was pulled off station in the South China Sea to support the recovery operation. She steamed to Johnston Island, spent the night, and by 0720 on Wednesday, 23 July 1969, slipped her mooring lines and departed Johnston Island en route to the Apollo 11 recovery area. *Arlington* was to provide communications support to *Hornet* and NASA during the recovery operation, as she had done previously during Frank Borman's December 1968 flight on Apollo 8. Borman had flown to the moon and orbited it while testing equipment and procedures which would later be employed during Apollo 11's flight. About the only thing that Borman did not do was to land on the moon. Additionally she hosted President Nixon and his immediate traveling party for the night of 23 July 1969. Nixon was a former U.S. Navy officer who had served during World War II. He was a supply officer and had visited Johnston Island twenty-five years earlier. But he had never before spent a night at sea on a Navy ship.

The Apollo 11 astronauts continued on their flight back to earth. They vacuumed the cabin of the CM to capture as much moon dust and potential pathogens as possible prior to splashdown. Despite their efforts, a fine layer of dust remained. During a final briefing for the Hornet Module Retrieval Team, NASA's team leader, Dr. Stullken, predicted that splashdown would occur at 0551 the next morning. He expected the entire recovery and retrieval operation to require no more than one hour, from the time that swimmers deployed after splashdown to the time when President Nixon began to talk with the astronauts on board ship. The best time by any previous team, to date, had been forty-six minutes.

Captain Seiberlich ordered the crew to freshen up the ship. A quarter of earth's population was going to see *Hornet* next day, and he wanted her looking sharp. BM2 Mickey N. Lowe was a boatswain's mate from Texas. He was placed in charge of a small crew of men painting "non-skid" on the quarterdeck of the ship. Mickey was just standing there, coffee cup in hand, watching his crew get the job done. He had just transferred to *Hornet* from a tour in Vietnam, where he saw action while assigned to river patrol boats at the Naval Support Activity DaNang. All of a sudden someone positioned just above his level test-fired the salute guns. It must have been the rancher from Montana, GMG3 William W. "Bill" Winchell, who was in charge of that crew. Mickey instinctively took cover and hit the deck, covering himself from head to foot in gray paint. His coffee mug did not survive.

ABH2 Gary P. Righter was in charge of a larger crew which was ordered to paint the entire flight deck during late afternoon, from bow to stern. Righter had a long line of men stretched across the deck from port to starboard, each armed with a paint roller and a bucket. That night, before the deck was fully dried, someone pulled a helicopter across the area where Marine One, the president's helicopter, was supposed to land and messed up the paint. Righter had his men paint the area over again and apply a fresh coat of light paint for the lines which encircled the aircraft on deck. He just knew the paint was not going to dry before Marine One landed on the deck early the next morning; he was afraid that it would skid on the slick deck and possibly fall over the side of the ship and into the drink. Righter would be blamed. Next morning, just as Marine One approached the ship, Righter purposefully left the area and sought refuge in the mess decks so he would not witness his own demise.

At 1400 three helicopters landed on *Hornet*, loaded with twenty-five reporters who were members of the White House press corps and who traveled with President Nixon throughout his overseas visit with foreign leaders. Their stopover on *Hornet*, to observe the splashdown of Apollo 11 and the president's welcoming ceremony, represented just the first phase of their trip.

JO2 Chan Cochran, the enlisted journalist from Ohio who would later meet Neil Armstrong in person during the latter's induction into the Ohio Hall of Fame, remembers well the entry made by members of the White House press corps into his Media Center on board *Hornet* that afternoon. A young Dan Rather burst into the office and demanded special treatment by announcing a list of things which he expected to receive—right now—for support. The XO, Commander Lamb, stood before him and allowed Rather to finish then calmly responded that he could have this and that, but that he would not be able to get that or the other thing. Rather ignored him and asked loudly for a phone so he could speak to the bridge, presumably with a view toward speaking directly with the CO of the ship. The OOD answered the phone, of course, and Rather proceeded to explain that "some flunky" down in the Media Center told him that he could not have such and such. The OOD asked Rather to name the officer who had denied Rather's requests. Rather read the XO's name off the nametag which was pinned to his chest. At that, the OOD informed Rather that Commander Lamb was the XO of the ship and that Commander Lamb's decisions were final.

Lt. Cdr. William G. Tasker was the ship's meteorologist. He recalled that a turf battle soon broke out between reporters from the White House press corps and those who had been on board for almost two weeks to cover the splashdown and recovery. Their battle was for precedence in filing of stories which were transmitted by the ship.

At 1500 USS *Goldsborough* (DDG-20) pulled astern of *Hornet* to serve as plane guard. *Goldsborough* would remain with *Hornet* throughout the recovery of Apollo 11 and the return of its astronauts to Pearl Harbor, and her crew was prepared to retrieve the CM in the event *Hornet* was, for any reason, unable to do so.

At 1600 Captain Seiberlich informed reporters on board *Hornet* at the splashdown point for the Apollo 11 recovery operation would be moved due to deteriorating weather conditions at the primary recovery zone. "We are all set to make recovery in the new zone," he stated.

Rear Admiral Davis's alert to NASA about deteriorating weather conditions had proven accurate. In response to the admiral's warning, NASA dispatched reconnaissance aircraft to the area and confirmed Captain Brandli's prediction that storm clouds would dominate the primary splashdown zone. Nearly thirty years would pass before information concerning the secret weather satellite program was declassified and the story about Captain Brandli's discovery and Cpt. Willard "Sam" Houston's decisive action to save Apollo 11 was revealed to the public. Captain Houston received a Navy Commendation Medal from the chief of naval operations in recognition of Houston's decisive action having contributed greatly toward the safe return of the Apollo 11 astronauts.

Thunderstorms were indeed moving through the primary zone. But the rain, wind, lightning, and heavy sea state bothered NASA less than the loss of visibility that deteriorating conditions created. Helicopter pilots simply had to be able to eyeball the CM as she descended by parachute toward the surface of the water. Failure to actually see the spacecraft could result in her loss in the endless expanse of waves which covered the Pacific Ocean and the possibility that the astronauts and their cargo of moon rocks would go to the bottom. Visibility, which ensured at least greater odds of success in getting to the downed spacecraft quickly, was a must. *Hornet* turned to the northeast. The new splashdown target area was 230 miles away.

Dr. Stullken intervened to calm the reporters. He said that NASA had informed the Apollo 11 astronauts of the change in splashdown points, and he added that the last-minute change would have no significant effect on the recovery operation. He noted with surety, "It really doesn't make a lot of difference what spot we're at out here in the middle of the ocean. Re-entry is the same. Splashdown will just happen a little later" ("Apollo 11 Briefs," *Idaho State Journal*, 24 July 1969, C-3). NASA calculated new re-entry coordinates and equations. *Hornet* began to make revolutions for a high-speed run to a position approximately nine hundred miles southwest of Honolulu. MM1 Enoust Lee's new bearing was put to the test and found to meet his captain's highest expectations, though the ship fairly shook with vibration as she cut through the waves at twenty-five knots.

At about 1700 Air Force One landed on Johnston Island on a direct flight from San Francisco. President Nixon and his travel party deplaned and almost immediately boarded Marine One for a flight to USS *Arlington*, which was already at sea and en route to the splashdown area. At 1736 Marine One landed on *Arlington* and the president took a brief tour of some portions of the ship. Nixon could be ill at ease and awkward during social interactions with some people, but he was right at home while talking to sailors. It was clear by his manner that he was excited about the prospect of witnessing the splashdown of Apollo 11 the next day, but he was genuinely kind toward and interested in speaking with the enlisted men on board *Arlington*.

The president planned to spend the night on board *Arlington*, then rise early to board Marine One for a short twenty-minute flight to *Hornet*. The last time a U.S. president visited a U.S. Navy ship named *Hornet* was exactly 104 years earlier, on 24 July 1865, when President Andrew Johnson visited the fifth U.S. Navy ship named *Hornet* during the Civil War. The aircraft carrier which President Nixon was scheduled to visit, USS *Hornet* (CVS-12), was the eighth such ship named *Hornet* in the history of the U.S. Navy.

The VIPs were gathering near. At 1813 Adm. John S. McCain Jr., CINCPAC, and Dr. Thomas O. Paine, the director of NASA, landed on *Hornet* and were welcomed aboard. They had flown to the ship from Pago Pago, American Samoa, on board a COD aircraft flown by *Hornet* pilot Lt. Cdr. Patrick L. "Red" Boyle, the V-2 Division officer (Catapults and Arresting Gear). Boyle had gone through flight school at the same time as Admiral McCain's son, John McCain, who was at that time a POW in Vietnam and would later serve as a U.S. Senator from Arizona.

The admiral addressed a gathering of *Hornet* crewmen on the flight deck shortly after deplaning. MM3 Clifford R. Burr, a machinist's mate from Texas, was among the crowd. He chuckled with the uncomfortable memory of Admiral McCain informing the men that it was just great to be with them that day "on board USS *Intrepid*." It was a slip. USS *Intrepid* (CVS-11) was an *Essex*-class aircraft carrier which was nearly identical to USS *Hornet* (CVS-12), and she was a ship to which Captain Seiberlich, CO of *Hornet*, had previously been assigned. Admiral McCain had probably done his homework on Captain Seiberlich prior to his arrival on board ship that morning, so the slip was understandable. But the rest of his presentation was inspiring. Lieutenant (jg) Beaulieu recalled that the admiral then gratefully acknowledged their hard work, to date, in preparation for the recovery of Apollo 11, and that Admiral McCain exhorted them to continue their dedication to the mission. The admiral mentioned in passing that he had fibbed about his age to get into the Navy as a sixteen-year-old kid. Through hard work and dedication to duty he managed to reach the rank of four-star admiral, and he believed that they could do the same thing if they worked for it (author interview of Peter D. Beaulieu, 05 September 2007).

Lt. William M. "Bill" Wasson was on duty on the bridge as the OOD when Admiral McCain arrived. He recalled that after McCain addressed members of the crew on the flight deck, he immediately entered the island and made his way to the bridge to shake hands with everyone standing watch. Wasson recalled that Admiral McCain was an old enlisted man himself, who appreciated the fact that while some might be able to fully enjoy the hoopla surrounding the splashdown of Apollo 11, others would miss the excitement because they had to work. Wasson was touched by the admiral's gesture, and he privately whispered to the diminutive officer that the admiral's son, John McCain—then a POW in North Vietnam—was in their prayers. McCain looked up into Wasson's eyes and thanked him, and that may have been the most poignant memory of Wasson's tour on USS *Hornet* (author interview of Bill Wasson, 5 February 2007).

Sometime during the day a most unusual telegram arrived. Alexander Volodin, the announcer for Voice of America who prepared to broadcast live the next day's festivities to millions of Russian citizens behind the Iron Curtain, translated the message for the crew. It read: "To the Cosmonauts of Apollo 11. Dear Lunamen, We are overwhelmed and very proud of your remarkable achievement." It was signed by the family of Murtazin, U.S.S.R., and had been transmitted from the village of Gomel in Byelorussia. (USS *Hornet* message, 23 July 1969: news release by Alexander Volodin, Voice of America [courtesy of Cpt. Chris W. Lamb, USN (ret)]). It was a most remarkable communication, given the frigid nature of Cold War relations between the United States and the Soviet Union at the time. Within that context, the telegram from Russia landed with a louder thud on *Hornet*'s deck that day than had the Soviet's Luna 15 probe when it crashed on the moon's surface just days earlier.

Later, during the early evening, following the memorable Russian communication, Lieutenant (jg) Beaulieu walked by the wardroom and overheard a NASA technician exclaim rather loudly into a phone, "What's that? The Comm satellite crapped out?" TACSAT, a tactical communications satellite, had failed. The TACSAT was the primary means by which the press pool and now, the White House press corps, filed their reports. It was also a single point of failure, meaning there was no backup communications satellite available in 1969. When the TACSAT went down, it went down, and that was that. By the worst of luck, it happened the night before the splashdown of Apollo 11. Two relay circuits were quickly established among *Hornet*, *Arlington*, and CTF-130 in Hawaii over the ATS-1 Goddard Space Flight Center satellite system, but were found to be of insufficient quality to support recovery operations. Communicators worked on alternate solutions throughout the night.

Fortunately members of the press pool had not learned about the failed satellite. They remained, for the most part, in the media center away from the wardroom and were not privy to what was going on. That changed, however. Lieutenant (jg) Powers, the young officer who was scheduled to serve as the ship's liaison officer and aide to President Nixon—who had just days earlier posed for photographs with his roommates while wearing the presidential gift ball caps—walked into the media center shortly after learning that the TACSAT had failed. He saw a communications officer enter the room and, not thinking, casually asked the officer whether he had been able to get communications back up. Big mistake. He had made his comment in the presence of reporters in the media center—men who were both hungry for a sensational story, perhaps a scoop, and who were dependent on communications to file their stories. Powers's question could have ignited a fire storm.

The communications officer quietly asked to speak with Powers privately in the passageway, outside the media center. When they were safely out of earshot of the reporters, the communications officer, a salty old Mustanger (an officer who came up through the enlisted ranks to attain his officer's commission the hard way), chewed out the young officer. "You must be the dumbest sailor on this ship!" he declared. "How could you say that in front of the press?" (author interview of Richard F. Powers, 2 April 2008). Powers was young and inexperienced in life, but he was learning fast, the hard way.

Elsewhere, UDT Detachment Apollo's officer in charge, Lt. Clancy Hatleberg, huddled in his stateroom with his fellow UDT officers, Wes

Chesser, Bob Rohrbach, and John McLachlan. They once again went over their plan for the recovery and generally discussed the situation. They could expect heavy seas and perhaps the worst overall weather conditions in which they had ever worked at sea. But they had proven just a week and half earlier during the worst storm in which they had swum to date, that they could get the job done. It was a pep talk. But none of them really needed to be shored up. They were UDT frogmen.

Chesser slipped away and found Commander Jones, the CO of HS-4 and pilot of the Recovery. Chesser believed that Jones would act as the tactical commander on the scene during the recovery. His decision to deploy swimmers would be relayed to the air boss and perhaps to *Hornet* for concurrence, but it was Jones who would make the call on the scene. So he was the man with whom Chesser needed to speak.

Wes Chesser was the leader of Swim TWO, the UDT swim team designated by Clancy Hatleberg as the primary recovery team within UDT Detachment Apollo. Chesser himself was the most experienced swimmer within the detachment in Apollo recovery operations, having served two tours of duty with the detachment and having directly participated in two previous, successful recoveries. Chesser was also the calm leader in whom Hatleberg had his greatest confidence; in a worst-case scenario— and it was beginning to look as though the recovery of Apollo 11 might occur during pretty rocky weather conditions—Chesser was the officer most capable of coolly assessing problems which occur while under stress and applying the correct solutions to resolve them, without fuss. Chesser was simply the right man for the job. But there was the matter of NASA's rule: The first swimmer helicopter to arrive on the splashdown site ordinarily deployed to execute the recovery. If Chesser's helicopter happened to arrive first on the scene, well, fine. He could expect to deploy with his team. If not, not.

Chesser did not want to engage Commander Jones in a protracted conversation or debate about the merits of this swim team over that swim team. To do so seemed improper. But he did believe, like Hatleberg, that Swim TWO had the best chance to pull off a successful recovery if conditions worsened. Clancy Hatleberg had already conveyed those thoughts to Commander Jones, to ensure that Jones was well equipped to make an informed decision on the morrow. Chesser's approach to Commander

Jones that night simply echoed an argument that Hatleberg had previously made.

Commander Jones, who would later rise through the ranks to the level of vice admiral of the Navy and wear three stars on his shoulder boards, exercised his best judgment of the situation that night and simply informed Chesser that he could not promise the young UDT officer anything. Chesser questioned, doubted, and cross-examined himself in private agony for decades over his action that night in approaching Commander Jones at all, with a plea to allow his team to execute the recovery of Apollo 11. In the end his interaction during the evening of 23 July 1969 did not matter. It simply was not a factor in Jones's final decision.

In a stateroom on board *Hornet* that night, one of Commander Jones's men sat down to write a letter to his daughter, Kimberly. Kimberly was less than two months old at the time, and Lt. Rich Barrett, the pilot of an HS-4 helicopter designated by call sign Swim TWO for the recovery operation, missed her greatly. In his letter Barrett expressed his hope that the letter would someday provide to Kimberly "a sense of history and a sense of pride in this great country of ours" (letter by Kimberly Barrett). Barrett sealed the letter inside a special USS *Hornet*–Apollo 11 recovery envelope and prepared to post it the next day, Splashdown day.

Later that night a final meeting was held in a ready room, with about thirty officers present, including Admiral McCain and Rear Admiral Davis. Captain Seiberlich delivered his final instructions. He reviewed the anticipated sequence of events for the recovery. He wanted to ensure that it was successful, and his message to his officers regarded the importance of not screwing it up. Lieutenant (jg) Beaulieu was in attendance for the meeting, and his understanding of Captain Seiberlich's message was summed up in a few simple phrases: This is important; everyone is watching; let's get this done right. His final word of advice added a nice touch to already fraying nerves: "If this thing starts to go to worms, just play it cool" (author interviews of Peter D. Beaulieu and Chris W. Lamb, 5 September 2007). Beaulieu would be on the front lines with his retrieval crew the next morning. He could only hope that *Hornet* would arrive in time.

A warship like *Hornet*, even an old ship that was at the end of her predicted life cycle, could easily turn revolutions fast enough to cover

230 miles overnight under ordinary circumstances. Of course *Hornet* was transiting through a spot with heavy sea state, which naturally slowed the ship. Still, it should not have been a problem, but they still did not have an exact fix on their location.

Hornet's navigators had been plagued for three days running with heavy cloud cover and, unable to see the stars to take a reading by sextant, they navigated by dead reckoning. In that zone north of the equator, drift amounted to about eight-tenths of a knot per hour, which could result in position error if miscalculated. *Hornet* had been drifting and circling and navigating by dead reckoning for three days, and God only knew whether their calculations were correct. The ship's navigator, the assistant navigator, QM1 Howard B. Mooney, and the XO stayed up all night, trying to find a break in the clouds through which to take some star sightings. No good. They convened in the navigation bridge and pre-calculated all of the data required for a firm fix, save star positions, so that their final calculations could be computed quickly, once their star sightings were made. They had only to find a break in cloud cover, which did not come.

One thing was certain, though, and that was that Apollo 11 would most likely make it to the new splashdown point on time and on target, in accordance with NASA's new calculations. They were that good. But whether *Hornet* would arrive in time to assist them was an open question.

12

The Big Day

THURSDAY, 24 JULY 1969

HOURS before the Apollo 11 astronauts entered earth's atmosphere to begin the final phase of their perilous flight, communicators on the ground in Houston radioed to inform them of recent and light-hearted events on earth in which they might have an interest.

President Nixon surprised your wives with a phone call from San Francisco just before he boarded a plane to fly out to meet you. . . . Air Canada says it has accepted 2,300 reservations for flights to the Moon in the past 5 days. It might be noted that more than 100 have been made by men for their mothers-in-law. . . . And finally, it appears that rather than killing romantic songs about the Moon, you have inspired hundreds of song writers. Nashville, Tennessee, which probably houses the largest collection of recording companies and song publishers in the country, now reports it is being flooded by Moon songs. Some will make it. (www.jsc. nasa.gov/history/mission_trans/AS11_PAO.pdf, accessed July 2006)

The astronauts had been much too busy, preparing for re-entry, to sleep. The youngest of the UDT frogmen, John Wolfram, reported that he did not get much sleep that night, either. He attributed his inability to sleep not to concern about the possibility that poor weather conditions

might await him in the water on Splashdown day—"We were professionals. We knew our jobs and had performed them hundreds of times. I
wasn't worried at all about messing up." Rather, his fitfulness that night
was attributable to a simple case of nervous energy. He wanted to get into
the water and get the mission behind him. He had been pent up on the
ship for almost two weeks, and he was tired of waiting. It was time to get
the game on (author interview of John M. Wolfram, 10 May 2007).

At 0200 UDT swimmers rose from their bunks and prepared for the
day. QM3 Mike Mallory, the strong workhorse of Swim TWO, documented the moment in his diary: "The BIG Day. Started at 0200. Went
down and had a little chow—then off to the briefing. About 0400 it was
off to the helos. Boy, do I ever have the cameras—5 in all—most of them
NASA."

During the briefing, swimmers were informed that weather and sea
conditions were not the best. Scattered clouds, but with ten miles of visibility, which was good for the helicopter pilots. Winds blew between
sixteen to twenty-four knots, and they could expect swells five to seven
feet. The wind could be a factor, insofar as it would push the command
module (CM) along at a good clip. The first swimmer would have to
capture the sea anchor ring on his first try.

Cdr. John S. Oster was in charge of air operations that day. He coordinated with a USAF liaison officer on board *Hornet* who maintained
communications with the two USAF HC-130 aircraft flown by the 41st
Aerospace Rescue and Recovery Wing. The officer reported that his Air
Force units were already airborne and positioned approximately 200–250
miles north and south of the new designated splashdown area. The team
was beginning to assemble. By 0400, the swimmers appeared topside on
the flight deck and prepared to board their helicopters. A small debate
ensued over the placement of their daisy.

Almost two months earlier, during the recovery of Apollo 10 on 26
May 1969, someone had slapped a colorful appliqué in the shape of a
daisy onto the nine-inch window of the Apollo 10 command module
just before it was hoisted on board the primary recovery ship (PRS) for
the Apollo 10 recovery, USS *Princeton* (LPH-5). Dr. Stullken, the NASA
team leader on board *Princeton* for that recovery, who was also known
among the enlisted swimmers as "Old Baldy," had taken exception to the
prank because the appliqué potentially marred the surface of the space-

craft's window and prevented NASA engineers from properly examining its condition after splashdown. Stullken did not know who the culprit was, so he chewed out the entire assemblage of crew and guests on board *Princeton* that day. Of course, the culprit was Mallory.

The swimmers were not fools. They understood that an attempt to slap a daisy on the CM of Apollo 11 in the presence of assembled VIPs, including the president of the United States, was not going to fly. But Stullken's coarsely worded threat was a challenge which simply required a response. With a daisy. Their first thought was to place one on the underbelly of Recovery. That was a very "visible" place for a daisy. Armstrong, Aldrin, and Collins could not fail to notice it as Recovery hoisted them from their spacecraft and into the hovering helicopter. They mentioned their plan to Hatleberg, their boss, who thought it was perhaps not the best idea he had heard all morning. But he wanted his men to remain loose, so he okayed it. Besides, he imagined the helicopter crewmen would spot the daisy before liftoff and remove it anyway. But the HS-4 guys beat them to it. They had already placed a sign on the underside of the Recovery to welcome the astronauts back home. It would be the first thing the astronauts of Apollo 11 saw as they were hoisted into the hovering helicopter before being whisked away toward *Hornet* for their welcoming ceremony. The swimmers would have to find their own spot.

One of the HS-4 crewmen was not in the best of moods that morning, anyway. He was among a small handful of men on *Hornet* who, frankly, were not happy about being there at that particular moment in time. Under the circumstances, ANH2 Richard B. Seaton could not be blamed. Seaton was previously assigned to another helicopter squadron, HS-8, and had just completed a tour in Vietnam when he transferred to HS-4. Seaton was married and had children at home. He no sooner transferred to HS-4 than the squadron deployed to USS *Hornet* for the recovery of Apollo 11. The practice sessions had been a lot of work and were not terribly challenging for him. He would rather have been home with his wife and kids. But duty called. He answered. And there he was. Interestingly, the passage of time, decades in fact, provided no different perspective for him. He understood that the flight of Apollo 11 was a historic event, and perhaps he should have been thrilled, even honored to play a role in the recovery. But he really wanted that time with his wife and kids after a long separation in Vietnam. He could not get it back.

Nearby, in a ready status at the catapult, Lt. Cdr. Patrick L. "Red" Boyle sat atop his C-1A Greyhound COD aircraft. A hatch in the overhead allowed the pilot to exit the aircraft and just sit on top, where he observed the activity around him. Boyle would sit there on the catapult throughout the day, bored and unable to participate in the festivities planned for the welcoming of the Apollo 11 astronauts. And yet, from the military's standpoint, Boyle played a critical role in the day's events. He was to stand by with his COD in the event NASA's decontamination procedure was violated. If "moon germs" somehow escaped into the atmosphere, and it appeared that the entire ship was about to be placed into a twenty-one-day at-sea quarantine, Boyle's orders were to whisk CINCPAC off the ship as quickly as possible. There was a war to prosecute in Vietnam, and Admiral McCain was additionally responsible for military operations throughout the rest of the Pacific theater. The nation could not afford to let him sit on a ship in the middle of nowhere for twenty-one days, in quarantine. Red Boyle sat on his COD aircraft all day long, prepared to launch at a moment's notice to ensure the admiral would not be stuck aboard ship until mid-August.

Men began to line up in the passageway outside of sick bay. They were "walking blood banks." USS *Hornet* men with blood types which matched those of the three Apollo 11 astronauts lined up in the event their blood might be needed by the astronauts in a medical emergency associated with the splashdown. Ten *Hornet* crewmen were designated to support each astronaut, and separate teams of doctors and other medical emergency personnel were also standing by in the event emergency medical treatment of the astronauts was required. Electrician's Mate 3 Gasho, the lighting electrician who borrowed another aircraft carrier's 24-inch carbon arc searchlight to support the recovery effort, was type B+, which was relatively rare. NASA thought of everything. The walking blood banks would remain in the passageway until a determination was made that their blood was not needed after all; in the process, of course, they lost their opportunity to observe the recovery itself. They were volunteers.

Lt. (jg) Richard F. Powers III woke early that morning, as well. He was understandably nervous. He would meet the president of the United States that day, in just a couple of hours, and he did not want to mess up. But as he looked around for the three presidential gift ball caps, which the president expected to present to the astronauts that morning,

Powers realized they were gone. They were not in the safe. In a panic, he returned to his room, hoping that he somehow had left them in there and forgot to lock them up the night before. When he stepped across the threshold to his stateroom, roommates Lauck and Wilson, the ship's public affairs officer, burst into laughter. They had hidden the ball caps as a prank. Powers retrieved the ball caps from them, but he did not think it was so funny.

By 0418, all aircraft involved in the recovery of Apollo 11 were in the air, including the five powerful HS-4 Sea King helicopters and two E-1B fixed-wing aircraft, Relay and Air Boss. The helicopters fanned out to their designated positions and maintained their distance and relative bearings from the ship by TACAN, the tactical navigation system. Every three minutes they checked in with *Hornet's* combat information center (CIC) to ensure they maintained an accurate position relative to the ship. They were figuratively welded to *Hornet's* hip; wherever *Hornet* went, they naturally followed.

The problem, of course, was that *Hornet* still did not quite know where she was. So everybody was off-target. Mission Control Center in Houston had no idea that *Hornet* was not yet in position. About an hour and a half before Apollo 11 was scheduled to enter the earth's atmosphere, Houston radioed Apollo 11: "The *Hornet* is on station just far enough off the target point to keep from getting hit. Recovery 1, or the chopper, is there; they're on station. And Hawaii Rescue 1 and 2, the C-130's, are within 40 minutes of your target point. Over" (www.jsc.nasa.gov/history/mission_trans/AS11_PAO.pdf, accessed July 2008). But neither Houston nor Armstrong nor the others knew that *Hornet* was not on station.

Captain Seiberlich retired to his stateroom late that night for some much-needed rest before Splashdown day. On one wall of the room hung a large banner which proclaimed *Hornet's* motto for the recovery of Apollo 11: Hornet + 3. That was his mission. But in fact, his ship was not "on station just far enough off the target point to keep from getting hit." Overcast skies continued to plague *Hornet* throughout her high-speed run to the new splashdown zone, so sextant readings were impossible.

The executive officer (XO) remained on watch throughout the night, along with the ship's navigator, the assistant navigator, and QM1 Howard

B. Mooney, manning the wings of the navigation bridge in case a break in the clouds offered the glimpse of starlight which would fix their position on the globe. The captain of *Hornet* was confident that his "Shitty-Little-Jobs Officer"—his XO—the detail man who actually ran the day-to-day operation of the ship, would remain on watch until the job was done. Those were the captain's orders, and Commander Lamb was well qualified to carry them out. For nearly eight months during *Hornet*'s final WestPac cruise, Commander Lamb served as the ship's navigator. He knew what he was doing. He also understood what was at stake.

Besides, the ship was not actually lost. Her navigator simply did not know exactly where on the globe she was positioned. There was a difference between the two situations. Men who sailed the oceans during the 1960s were used to navigation by sextant—and by dead reckoning when circumstances forced the latter option on them. Experience and training taught them how to deal with overcast skies, even when the seas were rough, a natural drift complicated their calculations, and a high-speed run was required. Their calculated, educated guesses—dead reckoning—could be relied on to keep them true enough to course to ensure they would arrive somewhere within the ballpark of their expected position for the splashdown. The real problem, Lamb understood, was that close proximity to the ballpark was not nearly good enough.

Hornet had her orders, and she was expected to proceed to a specific spot on the globe, and then to deploy assets to execute the Apollo 11 recovery operation. Failure to reach that spot on the globe, as ordered, was not acceptable insofar as it placed at greater risk the lives of American astronauts. Failure to eyeball the CM as she descended during hours of darkness, simply because *Hornet* was so far out of position that HS-4 pilots could not reach the spacecraft before splashdown, could lead to disaster. So Hornet simply had to make it, and make it on time. The Navy, NASA, and much of the world expected results, not excuses.

President Nixon's chief of staff, Bob Haldeman, spent a sleepless night on board *Arlington*, which steamed some twenty miles north of *Hornet*, over the horizon. Haldeman's cabin on *Arlington* was located next to the radio shack, and a banging door kept him awake all night. He rose at 0400 and soon emerged onto *Arlington*'s flight deck, which was absolutely dark. In short order the president's entourage gathered together, boarded Marine One, and at 0454 lifted off from *Arlington*'s flight deck. Hovering in wait, above them, two escort helicopters dispatched by *Hornet* took up positions near the president's aircraft to ensure that Marine One stayed on course en route to USS *Hornet*.

On the bridge of USS *Arlington*, Lt. (jg) K. W. Haber stood watch as the duty officer on deck (OOD). Haber made two entries into the official USS *Arlington* deck log concerning the departure of Marine One and the president's transit to *Hornet* to observe the recovery of Apollo 11. In both entries, Haber incorrectly identified *Hornet*'s hull designation as CVS-11 instead of the correct designation of CVS-12. Haber must have had Apollo 11 on his mind at the time.

On board *Hornet*, AC1 Anse E. Windham of Kingsville, Texas, was on duty as an air traffic controller within the Carrier Air Traffic Control Center (CATCC) that morning. His assignment to CATCC during Marine One's transit from *Arlington* to *Hornet* was not a coincidence. Windham was a competent, can-do air traffic controller who knew what he was doing.

The flight from *Arlington* to *Hornet* was but a short distance of twenty miles. But it was over open water during the black of night, and the Marine Corps pilot of Marine One was not used to navigating in those conditions. He called in to Windham for a position report at one-minute intervals. Windham recalled that the pilot seemed to drift about two degrees off course between calls. Had he traveled any distance at all, navigating by himself and without escort by *Hornet*'s helicopters, he would have been lost at sea. An aircraft carrier is a big target. But a darkened aircraft carrier at night, with running lights which appeared black against a black background, with no horizon at all, and with churning atop an ocean of blackness, was easy to miss. Windham kept the man on course.

ACC William "Bill" Wylie also worked in the CATCC. He recalled that when Marine One was "about eight minutes out," the radar screens on *Hornet* went blank. His worst fears had been realized. Earlier, the president's Secret Service advance team had padlocked for security purposes a porthole in the ship's transmitter room, a porthole which provided the only ventilation for the room. Wylie's objection at the time was based on fear that equipment in the room would overheat and shut down. Which it did. Fortunately an enterprising sailor positioned a bolt cutter nearby, just in case. The locks were removed, the room quickly vented, and *Hornet*'s radar screens flickered back into service. It was not a good day for the ship's radar to shut down.

Lieutenant (jg) Powers was seated in the wardroom at 0503 when he heard eight bongs sound over the 1MC and the announcement, "United States Arriving." "United States" was the call sign for the presidential aircraft, and Powers observed that a man could spend an entire career in the military and never get a chance to hear those words spoken over a loudspeaker. Powers knew that the announcement actually preceded touchdown of the president's helicopter on deck. A few more minutes would pass before he would set foot on deck. Powers awaited Nixon's arrival with some anxiety and trepidation. He was just a twenty-three-year-old kid, and he was about to meet the newly elected president of the United States and then serve as the president's official USS *Hornet* liaison officer and aide throughout his brief stay on board ship. Powers was sobered by the thought that "this was the varsity arriving"; serious business. He was nervous about meeting the president.

One minute later, at 0504, Hawaii Rescue ONE, one of the USAF HC-130 aircraft positioned more than one hundred miles beyond the designated splashdown zone in the event *Columbia* either undershot or overshot her splashdown, reported that he was on station. His U.S. Air Force para-rescue swimmers were positioned in his cargo hold with their aircraft deployed drift reduction system (ADDRS) and related rescue paraphernalia. A moment later, Hawaii Rescue TWO called *Hornet* to report that they, too, were in position.

At 0512, Marine One landed on *Hornet*'s flight deck. True to his word, ABH2 Gary P. Righter ducked inside the ship before splashdown and headed toward the mess deck, just in case the presidential helicopter skidded on the fresh coat of paint he had laid down on the flight deck just hours earlier. Still somewhat tacky, the paint apparently held, because Marine One settled onto the deck without incident.

Dawn was almost an hour away, so the president landed on *Hornet*'s deck in the dark of night. But Marine One's landing site was well lit, thanks to Electrician's Mates Third Class Gasho and Laurence. A crewman popped the hatch and lowered a set of stairs which led from the helicopter to the deck. At the bottom of those stairs stood young Lieutenant (jg) Ovard, the young officer who hoped some day to work in broadcast journalism. Ovard's job that day was to officially greet President Nixon as the president deplaned from Marine One, exchange salutes with him, and then step aside and remain available, and very much in the background, in the event any member of the president's entourage needed assistance of any kind.

Ovard was beyond nervous. He had fretted for days over this assignment because no one could tell him whether he should initiate the salute to the president, or whether he should wait for the president to salute, first and then respond with a salute of his own. And, when should he deliver his welcoming lines, before or after the salute? It was a simple thing. The correct answer, of course, was that a junior officer (Ovard) initiates the salute to a superior ranking officer (the president, as commander in chief), and that a greeting follows the salute. But Ovard, like so many of the junior officers on board *Hornet* that day—many of whom were assigned to perform somewhat intimidating assignments with heavy responsibility—was just a kid, and he was naturally nervous. As the last echoes of GMG3 Bill Winchell's twenty-one-gun salute faded in the wash of Marine One's rotors, Ovard managed a proper salute and issued his well-rehearsed, "Welcome aboard *Hornet*, Mr. President," before stepping aside.

The ship was enveloped in a shroud of darkness as it continued to push through the waves, but Electrician's Mate Third Class Gasho's lighting bathed Marine One in a bright circle of light on the flight deck. Gunner's Mate Third Class Winchell deployed a team of about thirty men to their shark watch posts around the ship. Each man was armed with a modern M-16 rifle. But new orders were issued to Winchell regarding ammunition: Full magazines were issued to each man, but shark watches were not permitted to lock and load until and unless they received a direct order to do so. The Navy wished to avoid accidental discharges.

At the bottom on the stairs, a bevy of VIPs greeted the president, including Admiral McCain, Rear Admiral Davis, and Captain Seiberlich. Dr. Thomas O. Paine was there, along with Frank Borman, commander of Apollo 8 and liaison officer for NASA to the White House. Further in the background was Lt. Henry Francis Dronzek, the administrative assistant to Captain Seiberlich who orchestrated the president's welcome ceremony. Secretary of State William P. Rogers next walked down the stairs, along with Chief of Staff Bob Haldeman and thirty-two-year-old Navy lieutenant commander Charles R. "Chuck" Larson, the president's naval aide who would serve the nation during future years as CINCPAC and then, twice, as superintendent of the U.S. Naval Academy in Annapolis, Maryland. Quite a crowd of men. Enlisted personnel scurried around the aircraft to chock its wheels and generally keep out of the way.

Had the president looked upward, he might have caught a glimpse of a handful of men standing pensively in the shadows of *Hornet's* island, on the wing of her navigation bridge. They were not peering down at the assemblage gathered round Marine One on the flight deck, however. They looked skyward, instead, still in search of a break in the clouds. The spacecraft of Apollo 11 continued to streak inexorably toward earth at a speed of approximately 25,000 mph. She could not stop or appreciably slow her own progress, and in just forty minutes she would end her flight with a resounding "whump" on the surface of the ocean, at precisely that point on the globe that was predicted by NASA's calculation. *Columbia* was on time, and on target. Hornet's XO still searched for a way to meet her.

A carpet was laid on the deck to form a path which the president could walk from the bottom of the stairs of Marine One and into the ship's island. So thorough were *Hornet's* preparations for Nixon's arrival that the ship even purchased a vacuum cleaner while in Pearl Harbor to ensure the carpet remained unsoiled for the welcoming ceremony—and then practiced vacuuming that carpet, and timed the practices, to ensure that a cleaning could be completed quickly, if necessary.

Unseen in the darkness, near the top of the O-7 level of *Hornet's* island, stood Lt. Cdr. William G. Klett, alone, microphone in hand. The commander of Task Force 130 and NASA ordered *Hornet* to provide a blow-by-blow narrative of events on board the ship as they unfolded that morning, and to broadcast a factually accurate account of events as they occurred, to a limited audience: the staff of TF-130 in Hawaii and NASA personnel manning the Mission Control Center in Houston. The other half of that broadcast team was Lt. Cdr. Clyde W. Jones, who assumed a position on the hangar deck starboard elevator where the module was to be recovered. Lieutenant Commander Klett was expected to report on events which occurred on the flight deck and beyond, including the splashdown and recovery operation, while Lieutenant Commander Jones covered the president's welcoming ceremony for the astronauts in the hangar bay after splashdown and the later retrieval of *Columbia* from the sea. Klett and Jones jokingly referred to themselves as *Hornet*'s version of Huntley and Brinkley, famous broadcast journalists of the day.

The president was escorted into the island and then down to the hangar bay for an initial orientation tour of the area where his later welcoming ceremony for the Apollo 11 astronauts would be held. Lieutenant (jg) Beaulieu observed that, ordinarily Dr. Stullken, chief of the thirty-five-member NASA recovery team on board *Hornet*, and a veteran of every single manned spacecraft recovery operation throughout NASA's history, would have hosted the tour for President Nixon and conducted his briefings. But Stullken elected to step aside and allow his assistant, John C. Stonesifer, to take the honors. Stonesifer had operated in Stullken's shadow for many years and had supported the chief of NASA's Flight Operations Division, throughout the manned spacecraft program. The honor of briefing the president of the United States was Stonesifer's reward for many years of faithful service. Lieutenant (jg) Beaulieu felt that that was a class act. The UDT swimmers might have been surprised to learn that the otherwise gruff Old Baldy had a soft side to him.

Hornet's hangar deck resembled a huge warehouse in its size and its spacious atmosphere, and it was filled with a variety of objects. One portion of the hangar deck, hangar bay #2, contained the two Mobile Quarantine Facilities (MQF), the dolly where Apollo 11's command module, *Columbia*, would rest after her retrieval from the water, and three ABC TV vans. That was the area in which President Nixon's televised welcoming ceremony for the astronauts was to take place. A length of tape on the deck served to mark the spot, right in front of the MQF's window, where the president would stand throughout the ceremony.

ABC had an exclusive contract to videotape the ceremony, and for that purpose fixed a couple of large studio cameras nearby. Camera angles caught not only the window of the MQF, in which the astronauts would appear during the ceremony, but also the ABC TV vans which were located immediately adjacent to the MQFs. Emblazoned on their sides was an ABC logo, which Captain Seiberlich deemed inappropriate.

The ceremony would be broadcast throughout the world, and the video-tape would be retained for posterity; Seiberlich did not wish to provide a venue for free advertising for ABC, so he ordered his crew to drape a large sign over the side of the nearest van, which simply read Apollo 11.

Lt. Shirley H. "Lee" Elliott recalled that a company placard had been mounted above the MQF's window by Airstream, Inc., of Jackson Center, Ohio, the manufacturer of the MQFs. Captain Seiberlich also took exception to the placard. "Airstream is not going to get all this free publicity," so he ordered his crew to cover the Airstream placard with one of his own, which reflected *Hornet*'s slogan for the Apollo 11 recovery: Hornet + 3. Lieutenant Elliott observed, however, that Airstream, Inc., did get at least one advertising plug, for a limited audience of three. But it seemed appropriate for three men whose quarantine inside the facility was driven by concerns that they might carry some kind of "bugs" back from their visit to the moon. Inside the MQF was a hand drawn sketch of Charlie Brown's cartoon dog, Snoopy, with a caption that read "Happiness is a bug free Air Stream."

Stonesifer briefed the president on the operation of the MQFs. He explained that the trailers included a kitchen equipped with something entirely new: a microwave oven, manufactured by Litton Industries. Microwave ovens were under development by major manufacturers during the late 1960s, but were not yet available for sale to consumers. The oven inside the MQF, then, was the very first operational microwave outside of a laboratory. A sufficient number of individual meal servings had been loaded by *Hornet* and stored in her freezers to feed the astronauts plus any other men who might be inadvertently contaminated during the recovery and retrieval process, for a period of at least twenty-one days, the length of a quarantine period specified by the Interagency Committee on Back Contamination (ICBC). NASA was at least somewhat familiar with the astronauts' food preferences, and had packed away a wide variety of menu choices to ensure their every desire was met. If everything went as planned, of course, the astronauts would be off-loaded in Hawaii in just two days, but NASA and *Hornet* were prepared in case their stay on board ship was extended. The trailers were maintained at a slight negative pressure to ensure that air easily flowed into them, and flowed outward only through filters designed to trap whatever moon pathogens might be present.

Stonesifer then drew the president's attention to a human mannequin clothed in a BIG suit. The man wearing the BIG suit that morning was twenty-three-year-old NASA engineer Brock Randall "Randy" Stone, a member of NASA's Landing and Recovery Division and a specialist in MQF operations. Like the trailers, the BIG suits were designed to enable the astronauts to inhale normally, but to exhale through a filter which entrapped lunar pathogens and prevented their escape into the outside atmosphere of earth. Upon completion of Stonesifer's explanation of the BIG suit, the president shook Randy Stone's hand. The orientation tour and familiarization briefing required but a few minutes, and by 0519 the president departed hangar bay #2 and followed his escort along a prearranged route into the ship's island and up to the flag bridge. He had been on board *Hornet* for only seven minutes, and splashdown was scheduled to occur in little more than half an hour. He had not yet had breakfast.

President Nixon's visit, of course, had been meticulously planned and choreographed. A U.S. Secret Service advance team had arrived days earlier to prepare for his visit and occupied a space within the flag plot while they worked. Agents reviewed the personnel file of every man on board the ship—more than 2,000 of them—and were frankly concerned about what they found in many of them. Lt. Cdr. Joseph A. Fidd, *Hornet's* supply officer, recalled a Secret Service agent stating that "*Hornet* might not be the best place for a president, because there were so many disciplinary cases on board." The Secret Service was so concerned, in fact, that they requested the Marine detachment (MARDET) lock a handful of *Hornet* crewmen in the brig until the president departed. To his credit, Captain Seiberlich refused to do so. But he did reach a compromise with the Secret Service. He agreed to confine the men to the sick bay, under guard, and that is what he did.

The president's route through the ship, from the time of his arrival until the moment of his departure, was examined, planned, and secured in advance. Hatches in certain passageways were dogged, and the passageways themselves guarded by U.S. Marines and members of the master at arms force. Word was passed to the crew that certain areas of the ship were simply off-limits throughout the president's stay, and they should plan accordingly. Secret Service agents were positioned in strategic areas. And the MARDET positioned a rapid response force on the hangar deck, just in case reinforcements might be needed to handle whatever emergency might occur. Commander Stuntz and Captain Caskey, the MARDET commander, stretched themselves almost to the limit in their effort to meet Secret Service requirements.

President Nixon arrived at the flag bridge, where he met Lieutenant (jg) Powers, the young officer who would serve as *Hornet*'s liaison and aide for him throughout the remainder of his stay on the ship. Powers had never met a famous person before, and he was very nervous. Nixon stepped forward, thrust his hand out and kicked off their relationship in a most congenial way, "I'm President Nixon. What's your name, Lieutenant, and where are you from?" Powers became so flustered that he lost his poise and actually forgot his name. He hesitated and stammered something, but the president was quick to put the young man at ease. Said Powers, many years later, quite humbly, "He could not have been more friendly and approachable." It took but a moment for Powers to recover, and Nixon filled the time by recalling his own days as a Navy officer and told Powers how excited he was to be on *Hornet* for such a historic event.

ACC Bill Wylie recalled that Captain Seiberlich went to great lengths to ensure that the flag bridge was in tip-top condition for the president's visit. *Hornet* enlisted men cleaned the area and applied a fresh coat of paint. The flag bridge was equipped with a chart, of course, which displayed the ship's position on the globe. Since the president was an old Navy officer himself, *Hornet* naturally anticipated that President Nixon would gravitate toward the chart to check on the ship's position. A dead-reckoning tracer (DRT) was affixed to that chart to display *Hornet*'s movements in the area, but *Hornet* crewmen realized with some chagrin that the DRT was inoperable. It lacked a motor to drive the tracer. That could prove embarrassing.

With the president of the United States about to visit, nothing less than perfection was acceptable. A quick search established that no spare DRT motors were available, however, so someone walked down below to Air Ops and cannibalized their DRT for a working motor. The objective was simply to create window dressing for the president, of course. Their commander in chief did not actually require a functioning DRT. But it was important to make a good impression on him, and equally important to avoid an embarrassing situation should he realize that *Hornet*'s operation in support of the recovery of Apollo 11 was not the picture of perfection which her commanding officer (CO) wished to portray. It was a small sacrifice for Air Ops to make. But the watch officer in Air Ops was absolutely livid. And that fact naturally amused his colleagues next door, in CATCC. They planned to rub salt into his wound later.

Next, the presidential breakfast. A great deal of effort had gone into the preparation of President Nixon's breakfast tray. The S-2, food services division chief, Lt. (jg) Raymond W. Dyer, was assigned to personally prepare the presidential doughnuts and pastries. The tray itself was procured during the stopover in Hawaii, along with white linens, which just fit a table brought up for the occasion from the first class petty officer's mess. And the ship's finest plated coffee urn was polished to a fare-thee-well. Captain Seiberlich released an official memo which detailed and choreographed every moment. Lt. Henry Francis Dronzek, the administrative assistant to Captain Seiberlich, brought the tray up from the flag mess. SDCS Charles Everett Raigans of Oklahoma, the senior steward on board *Hornet*, wearing black trousers, black cummerbund, and a gold jacket, presented the tray to the president in the flag plot. A card in script greeted the president quite formally, "Good morning, Mr. President. Your breakfast is served to you by SDCS Charles E. Raigans." And everything, except ruby red grapefruit—which was simply not to be had anywhere in the state of Hawaii for some reason—was tastefully arrayed on the tray. Taking no chances, Captain Seiberlich ensured that two identical presidential trays were prepared just in case someone tripped, but Nixon only had a cup of coffee. He was too excited by the events surrounding him to maintain an appetite.

Captain Seiberlich, who thought of everything and then actually prepared for it, had ordered the sewing of a special presidential Navy jacket in Richard Nixon's correct size, just in case the weather turned cool, or windy, or wet. The size was double-checked. And his name, plus the presidential seal (specially ordered), USS *Hornet*, and of course, Apollo 11 were sewn on the jacket in golden threads. Like breakfast, this was not needed either and Lamb ended up with it as a souvenir.

Lieutenant (jg) Powers, of course, walked onto the flag bridge along with everyone else, though he stood in the background. NASA director,

Dr. Thomas O. Paine explained to the president the sequence of events which were about to unfold. The secretary of state, the president's chief of staff, and others gathered round, gawking at the activity on *Hornet*'s four acres of flight deck below them.

Captain Seiberlich slipped away and made his way to the navigation bridge, one deck above the presidential party. There Commander Lamb informed the captain that they had had no luck so far. Clouds continued to cover the sky and prevented Lamb and the other men detailed from figuring out where on the globe their ship was currently fixed and from taking sightings with their sextants. *Hornet* was still in the ballpark, but if anything went wrong during the recovery heads would roll if others learned later that *Hornet* was out of position.

An animated and excited President Nixon continued to smile and gawk from the flag bridge's wing just beneath them, completely ignorant of the fact that *Hornet* and her array of planes and helicopters fanned out ahead of the ship were not quite sure where on earth they actually were at that moment. Splashdown was perhaps twenty minutes away.

Captain Seiberlich, of course, had a lot on his mind. But had he glanced just outside the navigation bridge, to the navigation bridge's wing which overlooked the flight deck, directly above President Nixon's position, he might have temporarily forgotten all about his troubles with sextants.

PH3 James Dewey and PH3 Paul Sanborn were members of USS *Hornet*'s photo lab and were posted that morning to the navigation bridge's wing to take whatever photographs might come their way. A Secret Service agent stood next to them.

PHCS Robert Lawson happened to be standing at that moment, directly beneath Dewey and Sanborn, on the flight deck some thirty feet below. Next to Lawson stood Captain Caskey, the MARDET commander. Lawson was the photo lab supervisor.

Dewey had a brand new Leica light meter in his shirt pocket. He leaned over the railing of the navigation wing to take a photo of the president, who stood directly beneath him not more than ten feet away. As Dewey leaned over the railing, his light meter naturally slipped out of his pocket and flew right past, and within inches of, Richard M. Nixon's face. Down below Lawson was watching his charges as they performed their duty. Captain Caskey was otherwise engaged. When the Dewey's light meter hit the deck, it smashed into a thousand pieces and sent Captain Caskey scrambling to figure out where the attack was coming from. Lawson, who saw the entire episode unfold, nearly had a heart attack.

Dewey was clueless. He was so focused on his job that he had not noticed a thing. But Sanborn had seen it fall. He leaned over toward Dewey and asked whether he still had his light meter. And the Secret Service agent intervened to suggest, "Don't let anything else fall" (author interview of James Dewey).

The mood behind Dewey was equally tense. Lt. Bill Wasson recalled that he was the "Recovery OOD" on duty during the operation to recover Apollo 11. He was selected for that position for his experience and demonstrated skill in handling the ship. At the moment, he was indeed steering the ship, and in that capacity he was well aware of the fact that his navigators were still feverishly trying to figure out where they were. Dr. Stullken, who stood near at hand, was also aware of *Hornet*'s predicament, and according to Wasson, Stullken "was beside himself with anxiety, anger, and frustration." Captain Seiberlich shuttled between the flag bridge and the navigation bridge, one deck apart, to keep up appearances down below and to exhort greater effort from above.

They finally caught their break. QM1 Howard B. Mooney remembers the exact time when he took his final sighting: It was 0526 or just twenty-five minutes before Apollo 11 was scheduled to hit the water. A small break in the clouds had given them the opportunity to sight on a sufficient number of stars to fix their position on the globe. Commander Lamb, the XO, sighted two stars. Lt. Cdr. Edward M. "Ned" Dunham, the assistant navigator, sighted one. And Mooney sighted the last one. That was enough. It was not the ideal number of sightings to firmly fix their exact position on the globe, but it was sufficient. While taking those star sightings and formulating their navigational calculations, the men had to consider the time of night when the sightings were made. That is how Mooney remembers, to this day, the time that he sighted the last star. It was an emotional moment.

And they were nine miles from the position on the plot which they had calculated by dead reckoning. That was not bad. In fact given that *Hornet* had been navigating by dead reckoning for three days, that she was naturally drifting above the equator at a rate of approximately eight-tenths of a mile per hour, and that she had just completed a high-speed run through rough seas for a distance of 230 miles, the accuracy of their dead reckoning calculations was remarkable. It was a fine demonstration of seamanship. Chris Lamb, the most experienced navigator among them, recalled that "it was amazing that we were so on-target." Still, they were out of position.

The Air Boss, the Relay aircraft, and all five of HS-4 helicopters had been keying off *Hornet*'s relative position since they had lifted off the ship, an hour and a half earlier. Every three minutes the pilots would check with the ship's CATCC to reconcile the TACAN readings with those of the ship, just to ensure that they remained in the desired configuration relative to the ship's position. But since the ship was out of position, then they were out of position as well. *Hornet* had to get moving, and quickly. Apollo 11 was scheduled to hit the outer shell of the earth's atmosphere at 0535—in nine minutes.

Hornet was making about fourteen knots at the time her navigators finally fixed her position firmly on the globe. There was simply no way in the world that she could crank up her engines and make enough revolutions to travel nine miles in nine minutes. But she had to do the best she could. Lieutenant Wasson called for all the steam in her boilers, flank speed, and, as Quarter Master First Class Mooney so indelicately put it many decades later, "Hornet began to haul ass." MM1 Enoust Lee's brand-new bearing was found to be up to the task.

Members of *Hornet*'s crew naturally noticed as she increased revolutions of her propellers and thereby picked up speed. Some of them assumed that she was "turning into the wind" to launch aircraft. Others were just enjoying the ride. It is not known whether President Nixon or any members of his party sensed that anything was wrong. A rumor later made its way among the crew that *Hornet*'s sudden increase in speed had something to do with the president, perhaps because he was in a hurry to get somewhere. But whatever the speculation regarding the ship's sudden increase in speed, they felt the vibrations of her steel hull pounding against the ocean swells, which caused the entire ship to shudder.

The president had returned to the flag bridge's wing to enjoy the show. Inside, some members of his party plus various hangers-on helped themselves to his untouched breakfast tray, much to the horror of Lieutenant Dronzek, the captain's administrative assistant. *Hornet* had prepared a separate tray for everyone else in the president's party, but it contained only a drab selection of doughnuts, freshly baked pastries, and coffee. The cannibalization of the president's tray was a disaster in the making, should the chief executive return to sample his Wheaties, so Dronzek whisked it away and out of sight.

At 0533 HS-4 helicopters Swim ONE and Swim TWO reported that they were positioned about ten miles from *Hornet*. Two minutes later, *Columbia* entered the earth's atmosphere. Right about that time Mike Mallory's parents arrived at London's Heathrow Airport, in the midst of an overseas trip. They were frantic to find a television set. His parents flagged down a porter and informed him that their son was a U.S. Navy frogman on board an American aircraft carrier, USS *Hornet,* in the Pacific Ocean at that very moment in time, and that he was involved in the effort to recover Apollo 11 after its splashdown in the ocean, which was just a minute away. The young man took Mike's parents into a porter's lounge at London Heathrow, introduced them to everyone in the room, and gave them a front row seat in front of their "telly." Mike Mallory was hovering at an altitude of some three thousand feet above the surface of the ocean, awaiting the first sighting of Apollo 11's fiery return to earth. Half a world away, near London, his parents settled into their seats to witness the event.

The spacecraft was traveling at a speed of approximately 25,000 miles per hour when she entered the atmosphere. Friction between the oxygenated air and the module's heat shield created a fireball that reached 5,000° Fahrenheit and effectively blocked all communication between Apollo 11 and her rescuers below for the next four minutes.

At 0539 and 0540, the Air Force HC-130 aircraft on patrol more than 100 miles distant reported sighting the fireball. At nearly the same time, within *Hornet*'s CIC, radar operators monitoring the ship's SPS-30 high altitude radar antennae reported contact at 230 degrees true, at a distance of 130 nautical miles from the ship.

Lt. Lee Elliott was on duty in the CIC when *Hornet*'s radar first returned a signal from the inbound Apollo spacecraft. He recalled a moment of alarm and confusion as the AN/SPS-30 long-range 3-D height-finding radar system initially provided a false range—also known as a second sweep echo—which indicated that the inbound CM was going to pass over *Hornet's* position and miss the anticipated splashdown point by about 250 miles. The AN/SPS-30 radar was not designed to process signals from a target which traveled at 25,000 miles per hour— threats to Navy aircraft carriers, for which the AN/SPS-30 was designed, were incapable of such high speeds—so the radar system produced inaccurate information to the CIC. That information was passed along to the bridge and must have caused some concern. But within seconds the AN/SPS-43A low-frequency, long-range air search radar system provided an accurate range, and the moment passed. *Hornet* began to produce accurate tracking information on the inbound CM.

Hornet posted a total of ten lookouts, with four on the O-7 level, the highest point on the ship's island, and another two forward, two aft, and two more at the port and starboard bows of the ship. Each lookout was teamed with a "talker" who maintained communications with the ship's CIC. Lookouts reported sighting a fireball at 210 degrees true. But men on board *Hornet* did not require binoculars to spot *Columbia* as she streaked toward her splashdown. "On the horizon, some 11 miles away, we saw what looked like a fireball trace its path against the early morning gray sky," described Lieutenant (jg) Powers, with the president on the flag bridge that morning. "It was an amazing sight," he recalled (author interview of Richard F. Powers, 2 April 2008).

Down below, on the flight deck, PHCS Robert L. Lawson had spent the previous seventy-two hours working nonstop to prepare for this moment. Lawson was in charge of *Hornet's* photo lab, and he had deployed enlisted photographer's mates all over the ship to document the day's activities. But he savored the moment that Neil Armstrong, Buzz Aldrin, and Mike Collins returned to earth in a blaze of fire. "It was truly awe inspiring," recalled Lawson, many decades after the event, "not in the sense that kids use that word today, but in its original context. It was awesome." For him and others who witnessed it, the images of that day still burned bright in their memories.

In the air, AWHC Stanley C. Robnett, the Recovery's hoist operator, and Lt. Clancy Hatleberg, officer in charge of UDT Detachment Apollo, shared a ringside seat to *Columbia's* blazing re-entry at the helicopter's

open doorway. They recalled that Recovery was in the dark at the time
that the fireball appeared, but the sky to the east was just beginning to
lighten with a reddish glow. They saw a huge streak of fire in the sky as
the CM entered the atmosphere on the horizon, but they lost sight of her
as the blaze died out and *Columbia* emerged from her blackout period.
That was Robnett's most vivid memory of the recovery operation, and he
would never forget it. Hatleberg, the rough-and-tough Navy commando,
was so moved by the memory that tears welled in his eyes while telling
that story. Decades after the event, he recalled that "it was one of the
most beautiful sights I'd ever seen" (author interview of Clancy Hatleberg, 19 August 2007).

On the bridge of USS *Hornet*, Lt. Bill Wasson, the Recovery OOD
who was steering the ship, dutifully made an official entry into the deck
log which was devoid of all emotion: "0541 The Apollo 11 Command
Module is sighted, bearing 240 position angle 3, making its re-entry to
the earth's atmosphere." HS-4 helicopters began their run uprange, in
the general direction of the CM. She could not be seen, of course, given
their distance from the spacecraft, and her emergency beacon had not
yet begun to transmit. But the sighting of the fireball had the effect of a
starter's pistol.

A minute later, at 0542, *Hornet*'s CIC reported that the ship's radar
had fixed the spacecraft's range at 65 nautical miles. *Columbia* was still
free-falling, but as she descended to an altitude of approximately 24,000
feet a barometric switch closed and set in motion a sequence of events
to slow her fall toward earth. Explosive devices blew the Apex cover off
the top of the spacecraft as planned, and mortars launched *Columbia*'s
drogue chutes skyward. She was still in the dark, but her emergency beacon was exposed and began to operate, and a flashing light provided a
visual signal to aid the helicopter pilots who were racing in her direction.
That was followed two minutes later at 0544 by reports of a double sonic
boom heard by *Hornet*'s lookouts. The astronauts had not yet established
communications with rescue forces arrayed down below, but *Columbia*
herself had spoken, and loudly. She had returned.

Lt. Lee Elliott, a native of Georgia who later settled in Tennessee, was
the first human being on earth to speak with the astronauts of Apollo 11
after they returned to earth. Until that moment no one knew whether

they had survived or succumbed to the heat of re-entry. Elliott was a watch officer in *Hornet*'s CIC, the focal point for operations. He called Apollo 11 several times before *Columbia*'s crew reported that they were in good shape and that they could see all three of the main parachutes fully deployed, plus the flashing light.

That was critical information for the men involved in the recovery operation. The report that Apollo 11's astronauts were in "good shape" meant that they had suffered no injuries and that they could, therefore, assist with their own recovery. The report that all three main parachutes had deployed was also very good news. The parachutes were designed to slow the craft's descent to a speed of little more than twenty mph upon impact with the surface of the ocean. Had only two of the three main parachutes fully deployed, the spacecraft would have descended at a faster rate of speed, with a much more jarring impact on splashdown— an impact which could produce injury. The astronauts were indeed fortunate that their earth landing system worked as designed. HS-4 rescue helicopters were too far distant from *Columbia* to see her in the dark as they sped in her general direction, but the flashing light should help them to spot her before she hit the water.

On board *Columbia*, the astronauts began to work through their checklists prior to splashdown. During his post-recovery debriefing, Apollo 11 astronaut Mike Collins, the commander of *Columbia*, recalled that "the big item for us was that we not contaminate the world by leaving the post-landing vent open. We had that underlined and circled in our procedures to close that vent valve prior to popping the circuit breakers on panel 250." He lightheartedly issued a recommendation for future astronauts, reminding them to "pay attention to that in their training: If you cut the power on panel 250 before you get the vent valve closed, in theory, the whole world gets contaminated, and everybody is mad at you" (http://history.nasa.gov/alsj/a11/a11tcdb.html#160, accessed July 2008).

On board *Hornet*, the immediate light show provided by *Columbia*'s fiery re-entry was over, and in the murky light of near dawn there was nothing more to see. Splashdown was less than ten minutes away, and *Hornet* was still out of position for that event, so her propellers continued to churn the sea at or near flank speed. But the show was over for the moment. It would be a while before *Hornet* could close the distance to the splashdown site. HS-4 and other air assets which were closer to the action would handle the next phase of the operation.

It might have been at that point that the president decided to duck back inside the flag bridge. Lieutenant (jg) Powers, his onboard liaison and aide, naturally followed. And that was the moment that Powers had been waiting for. He was alone with the president. If the president seemed excited earlier, Powers recalled, he was downright exuberant after witnessing the fireball and hearing the double sonic boom of *Columbia*'s re-entry. He was too worked up to sit, so he paced back and forth, alternately looking about the flag bridge and glancing out the window toward the flight deck.

Seiberlich's orders to Powers regarding his duty as liaison and aide to the president of the United States were quite clear. He was there to assist the president in the event the president required assistance. If the president required no assistance, then Powers was to remain passively in the background. He was not, under any circumstances, to initiate an interaction with the president of the United States. Powers was prohibited from engaging in friendly or familiar conversation with the president, unless the president initiated the conversation—even then, Powers was directed to be polite, respectful, brief, and professional. Still, he could not resist the temptation to request Nixon to autograph several presidential photos. Fortunately Nixon was in a great mood, and he was actually quite pleased by the request. He instantly agreed to sign the photographs and genially copied the script for each photo to ensure that Lauck and Wilson also received personalized gifts. "Hang on to those, son. They might be worth something some day," Nixon remarked as he handed back the signed photos.

At 0548, HS-4 helicopters Swim ONE and Swim TWO, along with the Recovery and *Hornet*'s CIC, reported contact with *Columbia*'s recovery beacon. She was transmitting. And that meant that HS-4 helicopters carrying UDT swim teams had found the signal with their onboard SARAH equipment and could zero in on the spacecraft's relative position with some accuracy. They had no idea how far away *Columbia* might be from their positions, but they knew which direction to take in order to find her. Lt. Richard J. Barrett, pilot of Swim TWO, recalled that he "put

Swim TWO into a dive [to quickly gain speed] and red-lined airspeed to get to splashdown as quickly and safely as possible."

John Wolfram, the youngest Navy commando, was on board Swim TWO. He recalled that his helicopter suddenly increased speed, and it seemed to him as though they were flying at two hundred miles per hour. The entire craft shook with the effort, and if not for the earphones on his head, the sound of engines and rotors would have deafened him forever. The race was really on. Swim ONE, which transported Lt. John McLachlan's team from UDT-12, was somewhere out there on the horizon, many miles away and not visible to Wolfram. The first helicopter to the splashdown site would deploy swimmers, and Wolfram wanted to get his feet wet that day.

By that time *Columbia* had descended to an altitude of 2,500 feet. She had not yet been sighted by HS-4 recovery forces—not yet eyeballed—but recovery forces were on her track and moving fast in her direction. Time was of the essence. To laypeople, 2,500 feet might seem like a great height, but for the astronauts of Apollo 11, it meant they had just minutes before splashdown. There was a better than fifty-fifty chance that *Columbia* would turn upside down shortly after splashdown, into a Stable 2 configuration until the astronauts righted her by inflating her flotation bags. During that time, her emergency radio beacon, her radio antennae, and flashing light would be underwater. Below, the sea was still in darkness, and *Columbia's* blackened heat shield, seared during the fiery re-entry, would blend into the sea and disappear. Swim ONE and Swim TWO were nearest her position and they raced to find her.

On the flag bridge's wing, the announcement that *Columbia* had passed an altitude of 2,500 feet as *Hornet* continued her high-speed run into the wind just added to the sense of excitement. The president leaned over the railing which overlooked the flight deck and peered into the coming dawn in an effort to see the spacecraft's parachutes among the clouds. He nudged his secretary of state and jokingly declared, "They'll probably land on the flight deck. Clear the flight deck!"

The mood was different on board Swim ONE. They were all business. Swim ONE was piloted by Cdr. Donald G. Richmond of HS-4, and co-piloted by Lt. William Wesley "Bill" Strawn. They transported the UDT-12 swimmer team that had joined UDT Detachment Apollo just weeks earlier. Swim ONE was led by Lt. (jg) John McLachlan of Spokane, Washington, and included PH2 Terry A. Muehlenbach of Chatsworth, California, and ADJ3 Mitchell L. "Mitch" Bucklew of Sanford, Florida (author interviews of Cdr. Donald G. Richmond, 5 August 2007; Mitchell L. Bucklew, 23 November 2008; John D. McLachlan, 1 August 2007; Terry A. Muehlenbach, 24 November 2008).

UDT Detachment Apollo's OIC, Lt. Clancy Hatleberg, and the rest of the UDT-11 frogmen in the detachment trained McLachlan's UDT-12 team in Apollo recovery procedures and found them equal to the task. But Hatleberg observed one relatively minor flaw in their performance which he could not easily dismiss: a tendency of one member of the team to suffer bouts of seasickness after spending hours atop the flotation collar or in a raft at sea during practice sessions. That was not an unheard-of occurrence, even among the toughest of frogmen. Swimming in the water for hours on end was one thing; any frogman could do that. But bobbing up and down while seated in a raft for long periods of time was quite another, and even the most solid performer occasionally became queasy in the stomach after a while. Relief was accomplished by vomiting. Hatleberg noted that one swimmer seemed to suffer more often than others.

Commander Richmond made an audiotape of his helicopter's communications that day and saved that recording for future reference. In the recording a technician seated in the body of the aircraft who monitored the SARAH equipment provided guidance to Commander Richmond as the craft raced toward the unseen Apollo 11 command module that was descending by parachute somewhere ahead of them. The technician's directions for Commander Richmond were issued in the calm and evenly spaced tones of a detached professional: "Left a little bit, sir. Good." "Straight ahead, sir." "To the right a little, sir." Listening to the audiotape, it is difficult to imagine that the aircraft was speeding along at or near to the Sea King's maximum speed of 166 mph as he spoke into the microphone.

Richmond and Strawn could not see the CM or its parachutes, but the Apollo 11 spacecraft's emergency beacon transmitted a signal which Swim ONE's SARAH equipment displayed to assist the pilots in their search. They were definitely flying along the correct bearing to intercept the CM. It was just a question of time. The signal emitted by the CM was a bit weaker than the signal usually transmitted by the boilerplate during practice sessions, a fact which a SARAH technician could interpret to mean that the CM was still some distance away.

Weather conditions at the splashdown site were considerably better than those at the previous site, but it was still quite cloudy. Richmond was flying in and out of clouds at an elevation of about eight hundred feet as he raced along the bearing indicated by his SARAH operator. He emerged suddenly from a cloud and screamed, "Oh, shit!" into his microphone as the window filled suddenly with the sight of three huge billowing parachutes, about one thousand feet directly in front of him. He almost hit the things, but veered away just in time to avoid a collision.

Commander Richmond recalled later that his exclamation of surprise at seeing the Apollo 11 command module suddenly appearing in front of him was "Oh, shit!" That may indeed have been in the officer's mind at the moment he realized that he was about to bring the flight of Apollo 11 to a sudden and quite tragic end. But his exclamation on tape was not quite as clear and articulate as "Oh, shit!" He sounded, there, more like a man swallowing a cat. Clearly, Commander Richmond was in distress.

The spread of parachutes above *Columbia*, each 83 feet in diameter, spanned perhaps 200 feet, or more. If Commander Richmond was indeed traveling at a rate of speed near to the Sea King's top speed of 166 mph and the distance from his helicopter to the CM was just 1,000 feet when Richmond first sighted it, then he had little more than four seconds of flying time to avoid a collision. The astronauts never saw him coming and likely first saw Swim ONE within their window only after Richmond swung around to the other side of the module.

Swim ONE radioed the contact to the Air Boss, who was operating at an elevation high above the action and then simply circled the descending module and followed her down to the surface of the water. Lt. (jg) John McLachlan's UDT-12 team was first on the scene. They prepared to deploy.

Splashdown occurred at 0549 local time, or 1249 in the afternoon in the nation's capital, at a position 950 miles southwest of Honolulu, Hawaii. *Columbia* was two minutes ahead of schedule. Given the fact that she traveled more than 240,000 miles from the moon, then made a last-minute course correction to reach a new splashdown site, NASA's calculations were amazingly on the mark. ADJ3 Mitchell L. Bucklew, a member of Swim ONE which circled the CM as it descended toward the surface of the ocean, snapped a photograph of *Columbia* just as it hit the water.

Mike Collins commented during a later debriefing that he felt a solid jolt at splashdown and that it was "a lot harder than I expected." *Columbia* hit the surface of the water at approximately twenty mph. The jarring impact so jostled the astronauts inside the spacecraft that they experienced a slight delay in cutting the three main parachutes loose from the craft. An eighteen-knot wind pulled the module over into a Stable 2 position—upside down in the water—amid four- to five-foot waves, and America's three returning heroes were happily though awkwardly suspended upside down in their harnesses, facing the ocean bottom. Commander Richmond, hovering above *Columbia* in Swim ONE, recalled that the parachutes dragged the spacecraft along the surface for a moment before the parachutes cut free.

The NASA recovery team had assigned a technician to serve as liaison to HS-4 throughout the recovery operation. That man's surname also happened to be Richmond—Mel Richmond—so the Swim ONE pilot and the NASA technician jokingly referred to themselves as cousins. Mel Richmond made clear to HS-4 that he hoped to recover the main parachutes for later examination, if possible. The parachutes typically remained near the surface of the water for only a couple of minutes before sinking, so deployment of swimmers to retrieve those parachutes would have to take place shortly after splashdown. But no order to deploy swimmers came from Commander Jones in the Recovery, so the parachutes began to sink and were soon lost.

Columbia's commander, Mike Collins, flipped three switches to activate the module's uprighting system, and air began to inflate three bags on the apex of the craft. USS *Hornet*'s deck log entry at 0550 was "Splashdown reported bearing 242, range 11 mi."

HS-4's Recovery arrived next and joined Swim ONE, which hovered above the module. Recovery reported that Columbia's dye marker was deployed, which spread a light green stain in the water. The main parachutes severed and floated just beneath the surface of the water, still visible. Commander Richmond and Lieutenant (jg) McLachlan snapped some 35mm photographs, and awaited the order to deploy the swimmers. AWHC Stanley C. Robnett, the hoist operator on Recovery, recalled that the sun was just coming up over the eastern horizon.

Deployment of swimmers was ordinarily ordered as soon as an Apollo module righted itself. The uprighting system required five to eight minutes to bring the craft into a Stable 1 or upright position, so Swim ONE and Recovery hovered nearby until the order to deploy could be issued. In the meantime, the astronauts took a second dose of medication, Dexedrine, to counter the effects of sea sickness. Wave action tossed them about quite roughly.

In his diary, Swim TWO member Mike Mallory observed that "the CM came down about 15 miles from the recovery ship—but only about 1½ miles from the *Arlington*. . . . Kind of odd the president flew over to the *Hornet* to see the CM come down, then it comes down right next to the other ship." It was true. *Arlington* was nearby. *Hornet* was still over the horizon. She was clearly out of position at the moment of splashdown, but, fortunately, she was close enough for HS-4 pilots to get a fix on the CM's emergency beacon before splashdown occurred.

Photo One, the HS-4 helicopter which transported official photographers to the splashdown site, arrived next. It was 0554, or some five minutes after splashdown. Photo One joined Swim ONE and Recovery over the uprighting CM.

JO2 Chauncey Cochran recalled that a quartermaster first class petty officer standing on *Hornet*'s island was the first man on the ship to spot the helicopters hovering above *Columbia* in the distance. He yelled it out "really loud," Cochran stated, "and a spontaneous cheer rose from crewmen gathered on the flight deck" (author interview of Chauncey Cochran, 6 February 2007).

Lt. Leroy Henry "Roy" Knaub, a COD pilot, recalled that, almost simultaneously President Nixon pulled the shirtsleeve of the man standing next to him and shouted, "There it is!" Knaub recalled that several men on the flag wing with President Nixon were equipped with binoculars. But the president was first among them to spot the helicopters, despite the fact that he had handed his own binoculars to another man and peered toward the horizon unaided.

Mallory and Swim TWO were not yet on the scene but they were en route. John Wolfram recalled that his aircraft fairly shook with vibration as it raced to get to the splashdown site.

On board Swim ONE, Lieutenant (jg) McLachlan's sea anchor swimmer stood in the doorway of his aircraft, prepared to jump into the water next to *Columbia* as soon as the order was given. The craft's uprighting system had almost completed its work. The three inflation bags mounted on the apex of *Columbia* were clear of the surface, and the craft canted at an approximate angle of 60°, nearly upright. At 0556, *Hornet* recorded that the Apollo 11 command module had reached a Stable 1 configuration.

But a conversation then taking place on board HS-4's Recovery stayed his deployment. As Recovery, Swim ONE, and Photo One hovered above the uprighting command module of Apollo 11, Lt. Clancy Hatleberg, the BIG swimmer and OIC of UDT Detachment Apollo, on board the Recovery and Cdr. Donald S. Jones, pilot of the Recovery and CO of HS-4, engaged in a brief exchange.

The details of that conversation were known to no more than four or five people at the time and not disclosed to others for another thirty-eight years after the recovery of Apollo 11. The upshot of the exchange meant that John McLachlan's UDT-12 swim team on board Swim ONE would not get their feet wet that day, despite the fact their team was first to arrive on the splashdown site, and contrary to NASA and HS-4 operational guidance and common procedures. The first swim team to arrive at the splashdown site ordinarily deployed to execute the recovery. That guiding principle meant that McLachlan's team, Swim ONE, should have expected to deploy to recover Apollo 11. But it did not do so. McLachlan, Muehlenbach, and Bucklew remained on board their helicopter, feet dry, and watched from a hover position above the action. John McLachlan was a good troop; he saluted the decision and did not complain. But that is not to say the decision did not rankle.

Hatleberg argued in favor of the deployment of Swim TWO, which had not yet arrived on the splashdown site, but was en route. His reasoning followed along these lines. The Navy could take no chances on that day. They had to deploy their best swim team to execute the recovery of Apollo 11. With the president of the United States on board the primary recovery ship, along with CINCPAC and one other flag officer, and with one-fourth of humanity tuning in to see the three American heroes who just returned from the very first walk on the moon, Lt. Clancy Hatleberg argued to belay the deployment of McLachlan's Swim ONE and instead wait for the arrival of his best swim team, Swim TWO. He felt it most advisable to deploy Swim TWO to execute the recovery of Apollo 11. Swim ONE would have to take a backseat and remain in the area as a backup team.

Swim TWO was the three-man team designated by Hatleberg as his detachment's primary recovery team; it was the most experienced recovery team then fielded by UDT Detachment Apollo. The team was under the command of Lt. (jg) Wesley T. Chesser, the most experienced Apollo recovery swimmer then in the detachment, and a man who earned Hatleberg's confidence by his ability to operate calmly under pressure. Its sea anchor man, John Wolfram, was hands-down the fastest swimmer in the detachment. And the third member of Swim TWO, Mike Mallory, was "just a horse in the water"; the guy was a powerhouse who could do anything. Besides, Hatleberg observed, there was no rush to deploy swimmers. Swimmers could not deploy until *Columbia* reached a Stable 1, upright configuration in the water, anyway. Waiting for the arrival of Swim TWO on the scene would not endanger the astronauts.

The final, though admittedly secondary, consideration was offered by Hatleberg during his brief exchange with Commander Jones, the on-scene commander: One member of Swim ONE had the misfortune to suffer too often from sea sickness, a malady commonly relieved by "puking over the side of his raft." Hatleberg's concern: the possibility that a global audience would conclude from such a display that moon germs had infected the swimmers, who operated in close proximity to the CM while installing a flotation collar and so forth, and that a plague was about to spread throughout the world's oceans.

The very likelihood of back contamination infecting humankind was laughable to many, possibly even Hatleberg. When reporters asked him about the possibility that moon germs might escape from the module when he first opened the hatch to toss BIG suits in to the Apollo 11 astronauts, Hatleberg cracked them up with a slow-motion pantomime, demonstrating the technique he planned to employ to snatch the little buggers one by one as they attempted their escape from *Columbia's* interior. Hatleberg had clearly demonstrated a sense of humor about the threat posed by moon germs. When reporters asked him how his wife might feel if he were contaminated and had to spend twenty-one days in quarantine with the astronauts inside their MQF, Hatleberg dead-panned, "I'd probably have to talk with my wife's lawyer about that." But he understood that perception carries the weight of fact. If the public perceived that moon germs caused a plague to spread, then no amount of calm, detached reasoning would dissuade them. Panic could truly result. And at a minimum, NASA and the U.S. government would have a huge public relations issue to resolve.

Hatleberg was likely more sensitive to the issue than other swimmers. He was the BIG swimmer after all, specially trained to execute decon-tamination procedures to cleanse both the astronauts and their CM. The fact that he spent more time dealing with the issue than other swim-mers suggested that he developed a heightened awareness of the threat posed by back contamination. But more to the point, Hatleberg engaged in conversation with Dr. Stullken, chief of the NASA recovery team on board *Hornet*. During those conversations, Dr. Stullken emphasized the importance of following decontamination procedures and emphasized, too, that failure to do so would automatically trip a quarantine proto-col which would confine *Hornet* and all men aboard her to a period of twenty-one days of quarantine at sea—and if that were to happen, said Stullken, "Don't even come back to the ship. You might as well start swimming for Australia." His message was clear: Don't mess up (author interviews of Clancy Hatleberg and Dr. Donald E. Stullken, 19 August 2007).

Hatleberg could just imagine one of his swimmers on camera before one-fourth of humanity, throwing up over the side of his raft. The malady might indeed be nothing more serious than seasickness, but it seemed doubtful that the worldwide audience would accept that explanation; more likely they would conclude that moon germs were killing the swim-mers. And Hatleberg simply did not want to take that chance.

Hatleberg's first line of reasoning favored the deployment of Swim TWO, his best team. His second line of reasoning, though secondary in his own mind, argued against the deployment of Swim ONE. Com-mander Jones was convinced. He radioed the Air Boss and sought concurrence with the decision from Dr. Stullken and the CO of *Hornet*. By the time their concurrence arrived, so too had Swim TWO. Lt. Richard J. Barrett, pilot of Swim TWO, was not privy to any of the discussion between Hatleberg, Commander Jones, or the Air Boss. Barrett recalled that, on arrival of Swim TWO, Commander Jones simply gave him the order to execute the recovery.

USS *Hornet*'s after-action report documented that at 0558, or just nine minutes after splashdown, the first swimmer entered the water to recover Apollo 11. QM3 Mike Mallory recorded into his diary, "John Wolfram was first out, with two sea anchors. He almost hit the CM when he jumped—as he came up out of the water he put his hand on it."

The public affairs officer on USS *Hornet* seemed confused as he related the action to press pool reporters on board the ship. He understood that Swim ONE arrived first on the scene after splashdown and naturally assumed that Swim ONE would deploy to execute the recovery. The report from Air Boss that Swim TWO just deployed a sea anchor swimmer caused *Hornet*'s public affairs officer to stammer just a bit.

A cameraman on board Photo One recorded the recovery, almost in its entirety. Wolfram, the fast swimmer from Wisconsin, did indeed jump as his helicopter made its first pass from downwind and nearly hit the CM. He advised years later that he did so because the wind was pushing the module along at a good rate of speed, and he wished to ensure that the module did not simply pass him by as he surfaced. He carried a spare sea anchor "just in case" the tether on the first sea anchor separated and in case a second sea anchor might be needed to slow the craft. Swimmers carried with them a device which enabled them to attach two sea anchors.

His prior experiences with sharks left the young swimmer a bit shark shy. He would not have admitted it during his youth, but several decades after the recovery occurred, Wolfram confided that the first thing he always did on entering the water was to look quickly around for sharks. He was alone in the water with them, with nothing to scramble onto out of harm's way in the event sharks challenged him, so he was naturally wary, if not a bit afraid. But he saw none after entering the water, so he quickly dismissed the thought. He did touch the CM as he surfaced, and, he observed many years later, it was very warm to the touch. "There are still some people out there who believe that the entire moon walk by Neil Armstrong was staged and that it never happened. But I can tell you, when *Columbia* burned through the atmosphere in a fireball—I saw it. And then she splashed down in the ocean, many minutes later—and she was still hot to the touch. I have news for those people: The astronauts really did it. This was the real deal" (author interview of John M. Wolfram).

Wolfram's first assignment was to check on the status of the astronauts. Armstrong, Aldrin, and Collins were again in communication by radio with the Air Boss overhead and with *Hornet*, but the sea anchor swimmer was required nevertheless to make a visual check on their status. At the water surface level, and on either side of the side crew hatch, were small portals. Wolfram looked inside one of those portals and received a thumbs-up signal from one of the astronauts. He assumed that it was Buzz Aldrin, but he could not be certain. John Wolfram, the youngest of the swimmers at age twenty, thus became the first person on earth to see the Apollo 11 astronauts on their return to earth. After receiving Aldrin's signal, Wolfram turned toward the Swim TWO helicopter hovering above him and gave the thumbs-up signal. By that time the CM, caught in the wind, began to flow past him. No more than a few seconds had passed since Wolfram entered the water.

Waves which were four to five feet in height rocked and turned the module. Wolfram was positioned on the upwind side of the module then, with Swim TWO circling around him for another pass while Photo One remained on the far side of the spacecraft, keeping clear of the action and recording it. A wave rocked *Columbia* downward toward the swimmer, but Wolfram's first attempt to grasp the recessed sea anchor ring failed as *Columbia* rocked backward and away from him. The craft moved away from him in the wind, and the young speedster gave a powerful kick to catch it and simultaneously grasp the ring which once again rocked downward in the wave action and toward him. That did it. Wolfram had it. But he was immediately and bodily pulled out of the water like a game fish as *Columbia* rocked in the other direction, and he hung on for dear life. Within seconds, though, he snapped one end of the sea anchor tether onto the ring and in one smooth motion reversed course to play it out and away from the craft. Less than two minutes had passed and the sea anchor was deployed. The craft visibly slowed as the anchor counter-acted the sail-like effect of the wind on the astronauts' vessel.

Swim TWO completed its circle and once again approached the slowed spacecraft from the downwind side. Trailing behind it was Wolfram's sea anchor, beneath the surface but quite visible from the helicopter's vantage point above. With the sea anchor now deployed and tethered just beneath the spacecraft's side crew hatch, the hatch itself remained oriented toward the upwind side. If the Apollo 11 astronauts had attempted to watch the action by peering out the main window on the side crew hatch, facing upwind, they would have seen nothing but wave tops and clouds, as Swim TWO circled round to the other, downwind side to deploy the next set of swimmers.

At exactly 0600, two minutes after the sea anchor swimmer, John Wolfram, had deployed, Lt. (jg) Wesley T. Chesser and QM3 Michael G. Mallory leaped from Swim TWO into the drink. Between them, they hauled the two-hundred-pound bag which contained the flotation collar. *Hornet* continued to vector toward the splashdown site at a speed of twenty-two knots, but she was just on or near the horizon and not at all visible from the surface level of the sea where the swimmers labored.

Among each team of swimmers, one man served the collateral duty as photographer. NASA and the press pool on board *Hornet* provided underwater cameras for their use. NASA wanted the swimmers to shoot specific parts of the CM prior to the hoisting operation, while the module was yet in relatively pristine condition. The press pool, of course, simply wanted juicy photos. On Swim TWO, Mike Mallory served that duty. He draped the four cameras around his neck, added one of his own to bring the total number to five, and leaped from HS-4 helicopter as it made a low and slow pass downwind from the CM—ten feet in altitude at approximately ten mph. One hand pulled the two-hundred-pound flotation collar bag out the door, while the other hand braced the mask on his face against the impact with the surface of the water. Somewhere along that route, he found a free hand to hold down the top of his scuba tank, to keep it from suddenly banging the back of his head on impact. Mike Mallory was just a horse in the water. He loved this stuff.

Despite the four- to five-foot waves, installation of the flotation collar went smoothly. "It went right on with no problem," Mallory later documented in his diary. Chesser, Mallory, and Wolfram worked as a team to install the flotation collar and encountered no kinks in the cables or other difficulties as they worked. It took them only three minutes to complete the job.

The very first time that NASA swimmers demonstrated for Mallory and Wolfram the "proper" technique to install a flotation collar onto a boilerplate, the NASA swimmers required thirty minutes to complete the job. And that was in a swimming pool. A quick review of NASA's Apollo Recovery Operational Procedures Manual—the written step-by-step procedures which governed every aspect of an Apollo recovery operation—revealed a flaw in NASA's cookbook procedures which simply did not jive with life in the real world of the open ocean. UDT swimmers corrected the error on their own.

The manual instructed swimmers to deploy from their helicopter on the downwind side of the CM, and to push the flotation collar bag to a point beneath the sea anchor ring and the side crew hatch of the spacecraft to begin the installation procedure. The problem, of course, was that once a sea anchor was deployed and tethered to that sea anchor ring, the drag on the ring oriented the side crew hatch toward the upwind side. In essence, NASA's cookbook instructed swimmers to push their 200-pound flotation collar bag from the downwind side of the CM, halfway around the floating spacecraft, until they reached the sea anchor ring, positioned directly beneath the side crew hatch on the upwind side of the spacecraft. That took some time and effort to accomplish.

Once there, NASA's instruction manual continued, swimmers were to snap one end of the collar's bungee cord to the sea anchor ring. One swimmer was to remain with the flotation collar bag, holding it snuggly in place beneath the side crew hatch, while one of the other swimmers swam all the way around the craft, pulling and tugging the flotation collar into place, with the assistance of the third swimmer. Having pulled the collar all the way around the craft, the swimmers would snap the other end of the bungee cord onto the sea anchor ring, inflate the collar by opening the valves on two CO_2 cartridges, and continue to snug the collar into a good fit as it inflated.

The procedure must have looked good on paper to engineers back in Houston, and seemed simple and straightforward in its theoretical execution. But in the water, it took a lot of time, and it wasted effort. Some bright UDT frogman realized during practice that, rather than push that 200-pound object halfway around the CM to the sea anchor ring, he could simply push the bag directly up to the downwind side of the module. There, a housing for the command module/service module (CM/SM) umbilical cables was positioned almost at the halfway point around the spacecraft from the sea anchor ring. Almost halfway, but not quite. Swimmers learned to push their flotation collar bag up and under that umbilical housing, and while one swimmer remained with the bag to keep it in proper position, the other two swimmers took either end of the bungee cord and swam in opposite directions around the CM until they met at the sea anchor ring to snap their ends of the bungee cord in place.

The fact that the CM/SM umbilical housing was not exactly at the halfway point on the downwind side of the spacecraft dictated that the distance around one side of the craft to the sea anchor ring was somewhat farther than the distance around the other side of the craft. One swimmer, then, found it easy to swim around the module to snap his bungee cord into place, while the other had a greater distance to pull his end of the bungee cord, and consequently a more difficult time to stretch the cord all the way to the sea anchor ring. Swimmers often combined their strength to pull and tug on that cord until it stretched all the way to the sea anchor ring. UDT swimmers thus revised and improved upon NASA's written instructions for installation of an Apollo flotation collar, and vastly improved the time required for installation. They were not rocket scientists. But they knew what they were doing in the open ocean.

The immediate affect the installation of the flotation collar produced on *Columbia* was to stabilize the craft within the wave action around it, much to the relief of astronauts inside. Swim TWO rounded the craft again and approached from downwind to drop a life raft into the water near the CM. Swimmers inflated the raft, then clipped it to the flotation collar immediately beneath the side crew hatch. That raft, designated raft #1, would be used by the astronauts during their upcoming egress and hoisting operation. Mike Mallory remained near the raft to ensure that it did not flip over in the downwash from the hovering helicopter's propellers. That was a very real problem. HS-4 and UDT swimmers experimented with a variety of methods to prevent prop blast from flipping over the rafts. Clancy Hatleberg recalled they once attached a handful of 5-pound weights to a raft, but when it was dropped from a Sea King helicopter into the water, the deflated raft sank like a rock. Swimmers resolved the issue by draping half a dozen "pockets" around the rafts, attached by strips of a newly developed space-age product called Velcro, which acted like mini sea anchors to catch the water and provide some drag when prop wash from hovering helicopters pulled on the rafts and threatened to flip them over.

Mallory also served as a safety swimmer, available to assist the astronauts in the unlikely event they were suddenly forced to perform an emergency egress from their module before preparations were completed for a normal egress. They were in good hands with Mallory nearby. Anyone who could swim unassisted for four miles as a youth, while both hands and feet were bound, could handle with ease any foundering astronaut who inexplicably slipped into the sea.

A minute later, by 0605, Swim TWO dropped a second raft into the water, designated raft #2, upwind from the CM. Chesser and Wolfram inflated the raft, then Wolfram ran a 70-foot tether from the raft to the CM's flotation collar. That raft remained upwind from the CM, for UDT frogman use, until the decontamination procedure was completed.

Hornet continued to steam in the general direction of the splashdown point at a speed of approximately twenty knots. At 0606, or seventeen minutes after splashdown, the MK56 gunfire control system in Radar 51, located on the starboard side of *Hornet,* just forward of the ship's island, locked onto the HS-4 helicopters hovering over the CM and provided range and bearing information to the Recovery OOD, Lt. Bill Wasson, on the navigation bridge. *Hornet* turned to intercept the floating module. Lt. (jg) Charles Stephen "Chuck" Rand, who served as the Officer in Charge (OIC) of Radar 51, provided range and bearing information to the bridge every minute thereafter, until 0627, after which range information only was passed along.

The Apollo 11 command module drifted along the surface of the ocean with the current, still rocked somewhat by wave action, but stabilized and buoyed by the flotation collar. With their rafts deployed, the frogmen of

Swim TWO were ready to begin the next phase of their recovery operation. HS-4's Recovery swung around upwind toward the swimmers' raft and dropped the BIG swimmer, Lt. Clancy Hatleberg, into the water.

Chesser and Wolfram awaited him below, while Mallory, still positioned near the CM, some seventy feet away, readied his cameras to take some photos. "As (Hatleberg) was dropped into the upwind raft I took a picture of the astronauts through their spacecraft windows."

Hatleberg joined Chesser and Wolfram in raft #2 and quickly pulled off his wetsuit top, then tossed his mask and fins aside. A couple of minutes later, the Recovery lowered his BIG suit, along with the "Bag O' BIGs" for the astronauts, plus some extra scuba tanks for the swimmers, and a tank each of Betadine and sodium hypochlorite. As they did so, the astronauts of Apollo 11 radioed the Air Boss, overhead, to advise, "All of us are excellent. Take your time." Mike Collins later related, "Nobody got sick. We each took a pill prior to entry and a second pill on the water. Those pills are called hyacinth and Dexedrine, and they seemed to work fine."

Chesser and Wolfram helped Hatleberg into his BIG suit. By that time, USS *Hornet* had closed to within seven miles of the CM and was visible on the horizon. It was 0613, and twenty-six minutes had passed since *Columbia* splashed down.

At 0615, the astronauts continued to communicate that they were fine and that there was no need to rush the recovery effort on their behalf. They reported that their spacecraft was "not as stable as *Hornet*, but stable enough." But Hatleberg was just about ready for them.

At 0618, Mallory lay down in raft #1, which was still fastened to the flotation collar, just outside and beneath the CM's side crew hatch. Mallory had a large and heavy scuba tank strapped to his back and still wore his mask, to avoid breathing stray moon germs in the area. Mallory single-handedly pulled raft #2, containing Hatleberg, Chesser, Wolfram, and their extra gear, the seventy-foot distance toward the CM and raft #1.

Hatleberg transferred from raft #2 to raft #1. The team transferred all of his decontamination gear to raft #1, as well, and then left him alone. Chesser entered the water, on scuba, and remained near the CM to serve as lifeguard during egress of the astronauts. Mallory and Wolfram, still aboard raft #2, then drifted to their original position, about seventy feet upwind of the CM, to avoid contamination. Hatleberg was on deck.

The BIG swimmer's first job was to ensure that vent valves located on the apex of the CM were closed. The astronauts were supposed to close those vents during their descent by parachute and prior to splashdown, but Hatleberg was supposed to double-check their work to ensure that lunar pathogens trapped inside the CM could not escape into earth's atmosphere at sea level.

Hatleberg was so concerned about the possibility that he might forget to check those vents, in his rush to get to the astronauts still sealed inside *Columbia*, that he wrote the word "vents" on the inside of his BIG suit mask as a reminder for himself. In his haste to do so, though, he wrote the word "vents" on the outside plate of his BIG suit mask, rather than on the inside. When Hatleberg put the mask on, the word "vents" appeared backward: "stnev." But it did the trick.

He remembered the requirement to check the vent valves, but he had a problem finding both of them. Hatleberg had trained on boiler-plates—mock-ups of operational Apollo CMs. And although the boiler-plates very closely resembled the operational models, the two were not entirely identical. Boilerplates did not have vents. The apex, or top of *Columbia* seemed to be crammed with a lot of unfamiliar equipment, jutting out here and there, so it took a moment, but Hatleberg did find one of the vents. He failed to find the other one, though. He moved on. Fortunately, the astronauts did indeed close those vents during their descent, and NASA's decontamination protocol remained intact.

At 0619, Hatleberg signaled the astronauts inside *Columbia* that he was ready to proceed to the next step in the process to egress the CM within the parameters developed by NASA, in coordination with ICBC, to minimize the possibility of contamination of the earth by moon germs. An astronaut opened the hatch for a moment, and Hatleberg tossed their Bag O' BIGs inside.

One might imagine that the hatch was opened for but the briefest moment of time, the better to ensure that air borne moon germs had little chance to escape the module. But as one watches the film of the recovery, recorded by a cameraman on board Photo One, it appears that Hatleberg fairly lounged in the hatchway for a moment after tossing the

Bag O' BIGs inside. There appeared to be no rush to close that hatch. Said Mike Collins, during his after-action debriefing by NASA, "Recovery operations went very smoothly. The swimmer threw the BIGs in to us. We put the BIGs on inside the spacecraft. We put them on in the lower equipment bay. Neil (Armstrong) did first, then I did, after him. Buzz put his on in the right-hand seat."

The astronauts required about eight minutes to don their BIG suits. During that time *Hornet* closed to within less than three miles of the module, making revolutions for twenty-one knots. Hatleberg busied himself by positioning his decontamination tanks, containing sodium hypochlorite and Betadine, near the side crew hatch, with the Betadine tank nearest to the hatch. He rested the tanks against the module and secured them by tucking the tanks snuggly beneath the bungee cord which wrapped around the module.

Collins recalled that the astronauts departed their floating spacecraft one at a time. First Armstrong, then Collins, and then Aldrin. Each wore a life vest which tucked under the armpit for flotation, and Hatleberg simply assisted them into a seated position in raft #1. The first astronaut to enter raft #1 seemed a bit animated and happy to be back; he shook Hatleberg's hand as he crossed the threshold and onto the flotation collar. As he settled into raft #1, he waved to the frogmen of Swim TWO, in raft #2, some seventy feet away.

Hatleberg recognized Mike Collins when the astronaut crossed the threshold while exiting the module. With the BIG suit on, the only part of Mike Collins's face which Hatleberg could see clearly were his eyes, but Hatleberg remembered those eyes from the time they spent together just weeks earlier during training in Houston. Collins must have recognized Hatleberg, as well, because he said something to Hatleberg as their eyes met. It sounded through the muffled BIG suit mask, like, "Mmmpph Um Ahhh Do MMph U!" Hatleberg was momentarily taken aback. As a young military officer, Hatleberg instinctively understood that when an officer of senior rank addresses an officer of junior rank, the junior officer was expected to respond. But Hatleberg had not understood a word of Collin's animated remarks. Thinking quickly, Hatleberg simply mimicked what he had heard from Collins, "Mmmpph Um Ahhh Do MMph U!" That seemed to satisfy the requirement. By 0629, or some forty minutes after they splashed down into the Pacific Ocean, all three astronauts were finally out of the confined space which they had called home for the last eight days.

The BIG swimmer then momentarily turned his back on the astronauts to complete his tasks. The first was to seal the hatch. But that did not work. Hatleberg shut the hatch and swung the handle around, but for some reason the handle did not ratchet downward to seal the hatch closed; rather, it swung freely, round and round. The locking mechanism seemed to be broken, and Hatleberg was at a loss.

That was a problem. Recovery and retrieval forces needed the hatch closed and sealed to ensure that seawater could not enter and flood the interior of the CM. Precious moon rocks were still stored on board *Columbia*. NASA wanted to examine the craft. And much of the world was watching. So a variety of factors favored the successful retrieval of the CM, not its sinking into the sea. *Hornet* was but a few miles off, then, slowing her approach from upwind, and she transported upward of one hundred members of the press. A sinking Apollo command module would make a sorry photo in the morning papers.

But that was hardly the worst-case scenario. The worst-case scenario was that the ICBC would declare the failure to secure the CM's hatch a violation of their decontamination protocols. If that happened, then everyone in the area, including everyone on board *Hornet*, would have to submit to a twenty-one-day quarantine at sea until a determination could be made regarding back contamination from the moon. And that message would broadcast on television and radio throughout the world, as well.

A contingency plan was in place to deal with the possibility that President Nixon might find himself in such a situation. Dr. Stullken's assistant, John C. Stonesifer, maintained direct communication via walkie-talkie with recovery forces then on station at the splashdown site. In the event that ICBC protocols appeared to have been violated and a twenty-one-day quarantine was about to be announced, Stonesifer arranged to signal President Nixon's Secret Service protective detail with a sharp salute. On receipt of that signal, the Secret Service would hustle the president onto Marine One for a quick departure. Presumably, CINCPAC would follow shortly thereafter on board Lt. Cdr. Red Boyle's C1-A Greyhound

COD aircraft, which stood ready on a *Hornet* catapult for just such an eventuality.

Hatleberg did not want to trigger that chain of events. Neil Armstrong rose from raft #1 to assist, but he had no luck. Mike Collins then replaced Armstrong at the hatch, checked to ensure that the hatch gear box was set on neutral and that the handle setting was also on neutral. It was not a terribly complicated piece of machinery, but it was more complicated than a car door. Collins found that the handle itself, which should have been locked into a restraining position, was just flopping free. He crammed it upward until it caught, and then the door could not only be closed, but sealed. Mike Collins was the CM commander, and it was his job to know the inner workings of every system in that craft. He was the right man for the job. Hatleberg would later recommend in his after-action report that the decontamination swimmer "be briefed on essential CM systems that he might have reason to function." To this day, Hatleberg still does not understand what went wrong with *Columbia*'s locking mechanism.

It took but a couple of minutes to seal the hatch. By 0631, the hatch was secured. *Hornet* was turning and slowing its approach from twenty-one knots to thirteen knots, upwind of the module. Hatleberg turned to the decontamination process. He turned his back momentarily on the astronauts and sprayed the apex of the CM with Betadine solution. The vent ports were up there; if lunar pathogens existed at all, then some of them might remain around those ports. Then he sprayed and swabbed all round the side crew hatch, to eliminate any pathogens which might have escaped during the astronauts' egress from the module. The decontamination solution made the flotation collar on which he stood quite slick, so maintaining balance and footing was a challenge for the young swimmer.

With that done, at 0634, Hatleberg turned back to the astronauts in raft #2, and began the process of swabbing them down with sodium hypochlorite. During decontamination training with Hatleberg in Houston, just weeks earlier, the BIG swimmer employed Betadine to swab each astronaut from head to foot, to kill lunar pathogens which adhered to their BIG suits during egress from the CM. Sometime after that training, NASA realized that Betadine tended to break down the waterproofing of the BIG suits, so a minor change in the decontamination procedure was made: a switch from Betadine to sodium hypochlorite as the chemical agent used to swab the astronauts and their BIG suits.

The Apollo 11 astronauts were informed of the change. But they were not aware of the problems encountered by UDT swimmers who played their roles during SIMEX training conducted by *Hornet* during the previous two weeks. Sodium hypochlorite emitted powerful fumes which burned the eyes and caused nausea. If astronauts failed to properly seal the hoods and masks of their BIG suits, vapors could enter the facial area and remain there, trapped. The reflexive response was to rip one's hood and mask off in order to breath fresh air, which was the last thing that the ICBC wanted to see. The very purpose of a BIG suit was to prevent lunar pathogens, which had possibly been inhaled by the Apollo 11 astronauts during their return from the moon, from escaping into earth's atmosphere. BIG suits did not protect the astronauts. In a sense, they were designed to protect earth *from* the astronauts and any pathogens which they might have inadvertently brought back from the moon. Their BIG suits were to remain sealed until the astronauts could be transferred into another sealed facility—the MQF, which was on board *Hornet*. Once inside the MQF, the astronauts' BIG suits could be removed. At that point, the MQF would serve to protect humankind (and most immediately, to protect the men on board *Hornet*) from lunar pathogens until the MQF could be transferred to a NASA laboratory in Houston, Texas, to complete the ICBC-mandated twenty-one-day quarantine period. The entire decontamination and quarantine process was designed to protect people from a deadly plague, and it was tightly choreographed and controlled. A violation of that protocol could mean, in the worst-case scenario, the spread of a deadly pathogen throughout the earth, for which no cure had been developed.

At a minimum, violation of the protocol would trigger a twenty-one-day quarantine requirement for everyone in the immediate area of the violation, to include *Hornet* and everyone on board her. And that is why caustic vapors emitted by sodium hypochlorite posed such a problem for the astronauts. If those vapors entered poorly fitted BIG suits, and if the astronauts then reflexively tore their hoods and masks off in order to reach fresh air, then deadly lunar pathogens might reach earth's atmosphere and trigger, at a minimum, the twenty-one-day quarantine requirement for *Hornet.*

A simple solution to the problem was developed by NASA officials on board *Hornet* during post-SIMEX testing. They instructed Hatleberg, the BIG swimmer, to minimize the amount of sodium hypochlorite solution used around the astronaut's facial area by squeezing excess solution from the mitt which he used to swab them down. The smaller quantity of solution produced a less concentrated vapor, and that procedure seemed to resolve the issue. But no one informed the astronauts of that small change in their decontamination procedure—squeezing the excess solution from the mitt prior to application onto their BIG suits—or of the problem encountered during Hornet's SIMEX which forced that change.

That oversight nearly caused a significant problem during the recovery. Mike Collins recalled during a NASA debriefing session,

> We sprayed one another down inside the raft. There was some confusion on the chemical agents. There were two bottles of chemical agents. One of them was Betadine, which is a soap-sudsy iodine solution, and the other one was sodium hypochlorite, a clear chemical spray. During our simulations [practiced sessions with Hatleberg weeks earlier, in Houston], we used Betadine in both bottles. They found that Betadine broke down the waterproofing in the (BIG) suit. They made a last-minute change and used Betadine for scrubbing down the spacecraft, but they used sodium hypochlorite for scrubbing us down. I had read about this and knew that there was a change. While the swimmer (Hatleberg) was scrubbing the spacecraft, I grabbed the other bottle and started scrubbing Neil (Armstrong) down. The swimmer got excited and didn't want me to do that. (http://history.nasa.gov/alsj/a11/a11tcdb/html#160, accessed July 2008)

Mike Collins and the other Apollo 11 astronauts were aware of the discovery that Betadine could compromise the waterproofing of their BIG suits, and that NASA directed the swimmers to instead use sodium hypochlorite, a bleaching agent, to swab down the astronauts. But the astronauts were not informed that the swimmers had discovered during their daily practice sessions in the Pacific that the application of too much sodium hypochlorite also caused a problem if the solution soaked through the BIG suits and permeated the cloth: it enabled noxious and caustic fumes to rise within the BIG suits and into the faceplates. The fumes of sodium hypochlorite could burn the eyes and throat, and force the astronauts to remove their faceplates—an action which would have violated the decontamination procedure and possibly forced the quarantine of the entire ship.

When Hatleberg turned around and saw Mike Collins spraying sodium hypochlorite onto Neil Armstrong, he instantly recognized the potential problem and intervened before sodium hypochlorite solution reached the intake valve on Armstrong's BIG suit. Had Collins sprayed sodium hypochlorite directly onto that valve, then Armstrong would have found himself in distress and might have pulled the hood and mask off his face to reach fresh air. If reporters then on board *Hornet* saw an astronaut in some kind of distress, they might conclude that moon germs, not the caustic fumes emitted by sodium hypochlorite, were the cause of Armstrong's distress. A report to that effect could have triggered a panic.

Collins and Armstrong were startled by Hatleberg's action and probably confused, as well, but in hindsight the twenty-five-year-old swimmer likely prevented a public relations problem from developing for NASA and the United States, at a minimum, by staying Mike Collins's arm during their "four men in a tub, rub-a-dub-dub" procedure. Mike Collins accurately summarized the lesson learned from the incident: "This is just another example where changes made between the training and the real thing have the potential of biting us."

All three astronauts were swabbed by Hatleberg with the mitt moistened with sodium hypochlorite, front and back, top to bottom. The astronauts then swabbed Hatleberg, in turn. The entire procedure required but eight minutes, and by 0642 Dr. Carpentier, the NASA physician then hovering above the CM in the HS-4 Recovery, reported to *Hornet* his judgment that there had been no breaks in the decontamination procedure.

A final scare remained etched into Clancy Hatleberg's memory for decades after the recovery: a rogue wave swept toward raft #1 and lifted it high into the air as it passed by. The astronauts must have seen it coming, because they held on for dear life. Hatleberg was amazed that no one was thrown overboard, including himself.

The commandos of Swim TWO remained throughout the decontamination process in raft #2, tethered seventy feet upwind of the CM to avoid windblown lunar pathogens which might escape from the immediate area. But with Hatleberg's decontamination process complete, the frogmen were free to move closer to the action. They did so at that time and took up positions in a semicircle around raft #1 to serve as lifeguards during the hoisting operation.

HS-4's Recovery, piloted by HS-4's commanding officer, Cdr. Donald S. Jones, moved into position directly above raft #1 to begin its hoisting operation. AWHC Stanley C. Robnett was the hoist operator; AWHC Nowell Wood operated a joystick to keep the helicopter perfectly positioned directly above the raft. Robnett lowered his Billy Pugh net to hoist the first of the Apollo 11 astronauts into Recovery. One hour after splashdown, the first of Apollo 11's astronauts was hoisted by Billy Pugh net into the Recovery. By then of course, the sun was fully above the horizon, it was broad daylight, and though the sky remained partly cloudy, the air was clear and visibility was unlimited. USS *Hornet* appeared close at hand.

Robnett had performed his hoisting operation on many occasions during practice sessions, of course. And he had served as the hoisting operator for the recovery of Apollo 10's astronauts, just weeks earlier. Like the astronauts, Robnett wore a BIG suit, and he recalled that it was extremely hot inside that suit during the recovery of Apollo 11. The outside temperature was in the mid-eighties. Beautiful weather. But the insulating facility of the BIG suit, combined with the outside temperature, plus heat buildup and retention inside the helicopter itself, and the natural buildup of body heat produced by physical exertion, created an uncomfortably hot environment. Robnett recalled that he sweated so much inside that BIG suit that, when he finally landed on *Hornet* a short while later and removed his BIG suit, he literally poured water, condensed sweat, from its facemask.

Robnett hoisted the astronauts in turn, then sat each into webbed seats and strapped safety belts over their laps. He was not able to speak with the astronauts, of course, given that all of them wore BIG suits and appeared for all the world like green bugs from outer space, but their animated body language clearly communicated their emotions at the moment. Said Robnett, "They acted like they were glad to be home."

The astronauts had been comfortable enough down below, in raft #1. But the heat buildup began to take its toll after boarding Recovery. Those who watched the recovery operation likely breathed a sigh of relief when they saw the astronauts safely aboard the helicopter and en route to *Hornet*. But relief is not what the astronauts of Apollo 11 felt at that time.

Mike Collins recalled,

> Aboard the helicopter, we started storing heat. For the first time, I became uncomfortably warm during the helicopter ride. That helicopter ride was as short as we are going to have them during this kind of operation. We debriefed the recovery people out on the ship and told them the same thing. When you get the [astronauts] on the helicopter, everybody shouldn't sit back and breathe a sigh of relief and think that the operation is all over; they should keep right on moving. This is the time when the (astronauts are) really starting to get uncomfortable. If the [astronauts] have to stay in that helicopter 15 or 20 minutes longer than we did, I guess the hood on the BIG would come off. That's a pretty wild guess. (http://history.nasa.gov/alsj/a11/a11tcdb/html#160, accessed July 2008)

During the hoisting operation, *Hornet* placed its whaleboat into the water to take up position approximately 150 feet from the CM. *Hornet's* retrieval team was already preparing to approach the module and hoist it on board. By 0653 that morning, all three of the Apollo 11 astronauts were on board the Recovery for their short flight to *Hornet*. A newspaper article in the *Sheboygan (Wisconsin) Press* noted that three Wisconsin Badgers played key roles in the operation: A young UDT frogman, John M. Wolfram, a native of Fort Atkinson, Wisconsin, served as the first swimmer in the water to recover Apollo 11 and the first human on earth to see the astronauts after their return from walking on the moon; Clancy Hatleberg, a native of Chippewa Falls, Wisconsin, served as the BIG swimmer and was the first man on earth to interact with the astronauts as they emerged from their spacecraft; and Cdr. Donald S. Jones, from Madison,

Wisconsin, was the pilot of the Recovery who flew the astronauts from their CM to the safety of USS *Hornet* nearby. To the folks in Sheboygan, Wisconsin, it must have seemed like a Badger recovery.

On board *Hornet*, Lt. Cdr. William G. Klett stood on the O-7 level with microphone in hand and provided a blow-by-blow description of the action in a broadcast back to TF-130 in Hawaii and to NASA's Mission Control Center in Houston, Texas. He informed his listeners that Recovery landed on the deck of USS *Hornet* at 0657, after a four-minute flight. Accompanying the Recovery was Swim ONE, in helicopter #53, which followed Recovery just in case it went down and required swimmer assistance. The HS-4 Photo One also accompanied Recovery back to the ship. Photo One landed a moment before Recovery to off-load still photographer Thomas M. "Milt" Putnam, whose duty at that moment was to take still photos as Recovery landed on deck.

At 0658, *Hornet*'s Recovery OOD, Lt. Bill Wasson, made his entry into the ship's deck log: "Recovered 2 helos. Astronauts N.A. Armstrong, LCOL M. Collins USAF, and COL E.E. Aldrin Jr. USAF." Lieutenant (jg) Beaulieu overheard R. Adm. Donald C. Davis remark, "Well, we got men all the way to the moon and back safely, and you just can't knock that!" Lt. (jg) Richard F. Powers III stood on the wing of the flag bridge with President Nixon as the Recovery landed on *Hornet*'s deck. The president was just beaming with pride, and some relief. The *Hornet* crew cheered as the helicopter touched down on the deck, and a twenty-eight-piece CINCPACFLT band played "Columbia, the Gem of the Ocean." Powers recalled that chills ran up and down his spine as the band then played "God Bless America."

Deckhands hooked Recovery up to a mule and towed it to an elevator to be lowered to the hangar bay. On arrival, the astronauts deplaned from the Recovery and, still garbed in their BIG suits, walked the short distance to the MQF. Crowds of sailors hailed them, and ABC recorded the event for posterity. A NASA official followed the astronauts and sprayed a disinfectant on the deck to kill any stray lunar pathogens which might have clung to the bottoms of their shoes. Dr. Bill Carpentier and MQF technician John Hirasaki joined the astronauts inside the MQF.

The official welcoming ceremony for the astronauts, hosted by President Nixon, would not take place for almost an hour. In the meantime,

Armstrong, Aldrin, and Collins disrobed from their awkward and by then uncomfortably warm BIG suits, took showers, and submitted to an initial medical examination by Dr. Carpentier. They were shielded from scrutiny by curious *Hornet* crewmen and others by curtains which covered the windows in the MQF.

Commander Jones, copilot Bill Strawn, and AWHC Nowell Wood deplaned from their Recovery; hoist operator AWHC Stanley C. Robnett remained on board to operate the brakes while Recovery was towed farther into the hangar bay after off-loading the astronauts. Robnett was then, finally, able to shed his BIG suit, pour the condensed water from his face mask, and join the crowd who stood by to watch the welcoming ceremony. The Recovery also had to be biologically isolated and decontaminated. The formaldehyde gas used for that purpose left an unpleasant residual odor which caused nausea among the crewmen working in and around the aircraft.

Recovery OOD Wasson steered the ship in a slow circle to make a downwind approach to the CM for the retrieval operation, scheduled to occur almost two hours later. As *Hornet* performed that maneuver, the frogmen began to clean up. Hatleberg disinfected raft #1, disrobed from his BIG suit, then used his knife to deflate the raft and sink it, in accordance with ICBC- and NASA-approved procedures. As QM3 Mike Mallory later recorded in his diary, "We sank the raft the Astros used to decontaminate in. So now the whole ocean is infected with moon germs." The tanks of disinfectant were drained, and then sunk, as well.

There was nothing for the swimmers to do, at that point, but wait for *Hornet* to complete her approach for the retrieval operation, which would occur some two hours later. That was a long time for a team of UDT frogmen to wait. During that time, they would miss the welcoming ceremony for the astronauts and would also miss an opportunity to see the president of the United States. They would miss all the fun. So they filled their downtime as only a team of Navy commandos might: by playing a long-running game of "King of the Command Module" on top of the Apollo 11 command module, the Navy's version of King of the Hill. Hatleberg stood atop the floating spacecraft and challenged all comers. He was not disappointed. Chesser, Wolfram, and Mallory charged up the sides of the module and did their best to pull Hatleberg down. In the end, as one might expect, the UDT detachment commander remained on top, as King of the Frogmen. The others would later claim that their attempts to unseat Hatleberg were half-hearted, at best. Had they been serious, the young lieutenant would have found himself in the drink. Rank had some, but not many, privileges after all, within the insular world of Navy special warfare. Inside the module, a two-hundred-pound cargo of lunar rocks and soil awaited its retrieval.

After it dropped still photographer Milt Putnam off on the carrier, Photo One returned to occasionally overfly *Columbia* and take photos. The young Navy frogmen waved in a friendly manner as the helicopter flew by. But they had tired of their game and had set about by then to pursue the more serious business of souvenir collection. Photo One unknowingly threatened to record their actions, so the swimmers had to be careful.

The exterior of *Columbia* was sheathed in kapton foil, a gold-leaf insulation material which absorbed UV radiation emitted by the sun and thereby protected the CM's occupants from deadly radiation. With the sun now fully in the sky, the CM looked like a golden pear floating on the sea. Small sections of foil had shredded during the fireball of re-entry to the earth's atmosphere, but the covering remained largely intact. Until the swimmers got to it, that is.

Kapton foil had the texture of a very lightweight aluminum foil. It crinkled and folded easily. And it readily peeled in sheets from the surface of the CM. It was therefore the perfect souvenir, providing, of course, the men on board Photo One did not catch swimmers in the act of removing it. So whenever Photo One made a pass overhead, the swimmers would wave in greeting. But the moment she banked to turn around, the swimmers would peel a length of kapton foil from the module, fold it, and stuff it inside their wetsuits for later retrieval. They were not foolish enough to remove a significant and therefore easily noticed portion of foil from the spacecraft. But they got enough. Decades later, a half-inch square of kapton foil from *Columbia* would sell on the Internet for $1,800. Mallory alone had a segment of foil which could easily cover a large picture frame. But it was a souvenir, for family and friends. Not for sale.

The nearly one-hour waiting period between the time that Recovery touched down on *Hornet*'s deck to the commencement of the official on-board welcoming ceremony for the astronauts proved difficult for the president to endure, as well. Lieutenant (jg) Powers recalled that President Nixon became increasingly anxious while waiting on the flag bridge for that time to pass. Captain Seiberlich had joined the presidential party on the bridge by then, and he did a masterful job of explaining to the president why he could not go down to the hangar bay to greet the astronauts right away, and why he would not be able to shake hands with them, at all.

Hornet's Huntley-Brinkley team as well as Lt. Cdr. William G. Klett and Lt. Cdr. Clyde W. Jones switched roles as the action moved from the skies above USS *Hornet* to the hangar bay. Lt. Lee Elliott joined Lieutenant Commander Jones in the hangar bay to provide commentary throughout the welcoming ceremony to listeners back in Hawaii and Houston, Texas.

Lt. Henry Francis Dronzek, the CO's executive assistant and acting master of ceremonies for the president's official on-board welcoming ceremony for the astronauts of Apollo 11, notified the bridge at 0750 that the astronauts were ready for the ceremony. The curtain was pulled away from the window in the MQF, and the three astronauts were seated just within that window, ball caps squarely in place on their heads.

Lieutenant (jg) Powers escorted the president down an escalator to the hangar bay. ACC Bill Wylie recalled that one of his airmen had been posted earlier that day at the bottom of the escalator to ensure for security purposes that no one approached or used the escalator before the president himself came by. He recalled that a man wearing a suit and tie attempted to board the escalator just before the ceremony, and that his young airman dutifully stood his ground and prevented the man from using the escalator. ACC Wylie observed that the man's attire, which was worn at sea in the Pacific Ocean, probably should have served as a clue to the young enlisted Navy man, but orders were orders, so the airman stopped the well-dressed man in his tracks, advising, "Sorry, sir, but I can't let you go up there." The stranger reached under his coat and displayed for the airman the stock of his automatic weapon. The Secret Service's memo had apparently not filtered down to the young sailor's level, but he took the hint and stepped aside. "Oh, I think you're okay to go up, sir," he said, and that ended the incident.

The president's route to and from hangar bay #2 had been secured hours earlier by Secret Service and *Hornet* security personnel. Passageways which led toward and intersected with the presidential route were secured and guarded by Secret Service agents and fourteen USMC guards from *Hornet*'s MARDET. An additional eight USMC enlisted men were stationed in hangar bay #3 as a ready reaction force in case they were needed while the president was meeting with the astronauts in hangar bay #2 nearby.

The president required instantaneous telephone access to the outside world while he was on board USS *Hornet*, regardless of where on the ship he might be at any given moment. At the top of the escalator, a telephone mounted on the bulkhead was specially wired to the White House situation room, just in case the president needed to take or place a phone call. Another awaited him in hangar bay #2, where he would meet with the astronauts. All he had to do was to pick it up and speak. It was not used, but the Secret Service tried to anticipate any contingency which might arise during President Nixon's three-hour visit on board USS *Hornet*.

The president arrived to the sounds of Ruffles and Flourishes, played by the CINCPACFLT band. Sgt. Rod Coyne, his U.S. Marine Corps orderly throughout the president's visit on board *Hornet*—reputedly the most squared-away member of the ship's Marine detachment—preceded him into the hangar bay, which was crammed with young sailors, all dressed in their whites. The bay itself was bathed in the glow of artificial light, in accordance with requirements levied on the Navy by ABC, the network which held an exclusive contract to videotape the ceremony live for viewers around the world.

The power requirement for the ceremony in hangar bay #2 was more than five times the requirement predicted during the 12 June 1969 planning conference for the recovery, held on board USS *Hornet* in Long Beach, California, but well below the capability of the ship's service turbo generators. Generating sufficient power for the ceremony was not a problem. Installation of the lights, however, had proven problematic: ABC lighting technicians were forced to work a single-shift, eight-hour workday during installation, because ABC was reluctant to authorize overtime pay.

AN Lee Bigbee of Milford, New Jersey, was a member of *Hornet*'s photo lab. His supervisor, PHCS Robert L. "Bob" Lawson, had directed members of his lab to position themselves at various spots throughout the ship to take official photos for a special USS *Hornet* commemorative book. Bigbee's assignment was to stay near and to photograph President Nixon while the chief executive was in hangar bay #2 with the astronauts. Bigbee was just an E-3, a very junior enlisted man, and his assignment was fraught with heavy responsibility. But as he maneuvered for space near the MQF to take his shots, a civilian pulled him aside and away from his duties, with the excuse that Navy personnel were not to approach within twenty-five feet of the president. The civilian, whom Bigbee assumed was an ABC employee, pulled Bigbee behind the platform which supported ABC TV's fixed camera for the event.

PH2 Roger J. Parolini was another member of *Hornet*'s photo lab who was assigned to take official photographs in hangar bay #2 that day. Parolini recalled that virtually everyone in the bay seemed to have a camera and that all of the professional photographers in the area were rubbing elbows, vying for the best angles for a shot.

ETR3 Dennis R. Moran found a great spot for camera shots, but it was short-lived. He was high above the ceremony, on a catwalk which traversed hangar bay #3, located immediately beneath the flight deck. Several enlisted men had taken roost on that catwalk to take pictures. It was a great location. But the Secret Service became alarmed by their unexpected presence and directed them to clear the area.

Lt. Henry Francis Dronzek, the master of ceremonies, called the crew to attention, and then to parade rest. President Nixon then asked the chaplain, Cdr. John A. Piirto, for a prayer of thanksgiving.

Lt. Lee Elliott, the CIC watch stander who first contacted Apollo 11 by radio as she descended by parachute toward the surface of the sea, was present during the ceremony and recalled that Commander Piirto's prayer seemed to go on and on forever. JO2 Michael R. Wheat estimated that the prayer lasted from eight to ten minutes and droned on so long that the crowd squirmed in frustration. ABC broadcast the first minute or so of Commander Piirto's prayer, but cut the remainder from subsequent broadcasts. Lt. Elliot later teased Commander Piirto about the duration of that prayer, saying, "John, that was the largest congregation you've ever addressed in your life. 500 million people. And it seemed like you just couldn't resist talking to them as long as possible"(author interview of

Shirley H. Elliot, 29 February 2008). But when he finished, Lieutenant Dronzek signaled for a drumroll, and the hangar bay was filled with the sound of the national anthem.

At 0755—more than two hours after the astronauts splashed down into the Pacific Ocean to fulfill President Kennedy's national goal of, first, placing a man on the moon before the decade was out, and then to bring him safely back home, and not quite an hour after they landed on the deck of USS *Hornet*—the president of the United States addressed the astronauts of Apollo 11. He thanked them, bantered awkwardly with them, and showered them with praise, as one might expect. "This is the greatest week in the history of the world since the creation," he effused, and invited them to dine with him at the White House on 13 August 1969, an invitation which the astronauts immediately accepted. Hundreds of flashbulbs flashed, and ABC cameras rolled to record the scene for posterity.

At 0803, just ten minutes after President Nixon arrived in hangar bay #2 to the sounds of Ruffles and Flourishes, he departed the area en route to the flight deck, where Marine One prepared for liftoff. As he stepped onto the flight deck, the president was greeted by a gathering of *Hornet* crewmen, some of them in whites, but many in the multicolored shirts of flight deck specialists who were working that day. To the consternation of some Secret Service personnel, Nixon waded into the crowd to shake hands and exchange remarks. The president was exuberant. He shook one man's hand and praised the men and ship for doing such a fine job. Excellence was to be expected. So the youngster replied quite matter-of-factly, "Yessir. We're the *Hornet*."

EM3 Michael Lee Laurence recalled that he shook hands with President Nixon on the flight deck that day. Then he turned to ham it up in front of an ABC video camera, shouting, "Hi, mom!" At home in Wyoming, his mom and dad watched the ceremony and did, indeed, see their son shaking the president's hand.

Lt. (jg) Robert I. "Bob" Abelson recalled that his parents, too, watched the coverage of the president's departure from USS *Hornet* that day, and they anxiously scanned the crowds of milling sailors on the flight deck in search for the face of their son. Spotting the familiar face, they jumped with excitement as Nixon shook hands with a young officer dressed in spotless whites before continuing in the direction of his helicopter. But Bob Abelson was nowhere near the president at that time: his parents were mistaken. Like so many other USS *Hornet* crewmen at that moment, Bob Abelson had work to do and no time to enjoy the festivities. He served as a watch officer in *Hornet*'s CIC, who relayed information from CIC to the bridge. For Bob Abelson and many of his fellow shipmates on board the aircraft carrier that day, it was just another day at work; just another day of duty at sea. Lt. (jg) James E. Ovard escorted the president to the helicopter and executed an official salute at his departure from the ship. The time was 0811, and the president had been on board *Hornet* for three hours.

ACı Anse E. Windham was still on duty within *Hornet*'s CATCC to guide Marine One during her departure from the ship. ACC Bill Wylie was on duty, as well. Wylie recalled that CATCC was equipped with an overhead speaker which enabled everyone within CATCC to hear communications with Marine One. Wylie recalled that the pilot of Marine One thanked *Hornet* for her hospitality, and just then, the Air Ops officer poked his head into CATCC to ask what that pilot had said. It was the same officer, Wylie recalled, who had complained ad nauseam days earlier when someone from the flag bridge "borrowed" the motor from Air Ops' DRT tracer to ensure the president would have an operable navigational chart on the flag bridge during his visit. Windham could not resist rubbing salt into the wound. When the Air Ops officer asked, "What did he say?" Windham glanced first at Wylie, and then responded: "He said thanks for the motors for the DRT."

Marine One flew to Johnston Island, where Air Force One waited to transport President Nixon to his first stop on a trip through East and South Asia. During that tour, he would stop in the Philippines; Indonesia; Vietnam, where a war still raged; Thailand; India; Pakistan; and, curiously, just before going home to the White House in Washington, D.C., to Romania.

The next day, on 25 July 1969, President Nixon issued a policy announcement concerning the nation's interactions with countries in Asia, which became known as the Nixon Doctrine. That policy limited America's military intervention in Asian conflicts and promised in the future to provide economic aid to anti-Communist movements, but to refrain from providing military aid to them. The announcement marked a significant shift in policy for a nation which had bled its youth during nearly a decade of warfare in Vietnam.

The president's triumphant celebration of the return of the Apollo 11 astronauts from the moon, combined with his new policy toward Asia, signaled to other nations in the world that the United States was determined to set foot on a new path. To lend teeth to his announcement, President Nixon dispatched national security adviser Henry Kissinger on a secret mission to Paris during his curious stopover in Romania, to meet with North Vietnamese officials for the first in a series of secret negotiations between the two countries which would lead in time to the end of America's involvement in the Vietnam War.

At 0830, Admiral McCain, CINCPAC, also departed the ship. His transportation was a COD aircraft flown by Lt. Cdr. Red Boyle. During the flight to Pago Pago, Admiral McCain approached the cockpit and asked Lieutenant Commander Boyle for a slip of paper. Boyle always kept some scratch paper and a pen in the knee pad of his flight suit. Moments later, the diminutive admiral handed Boyle a handwritten message for transmission to the men of USS *Hornet*, to commend them for a job well done. Lieutenant Commander Boyle retained that piece of paper as a souvenir. It remained, some four decades later, buried among other papers from that time, somewhere in a box, somewhere in either the attic or garage of his home. He is not sure exactly where the paperwork might be. But he does have the memory close at hand.

Lt. William M. Bill Wasson, the Recovery OOD, waited until both the president and CINCPAC departed before starting the approach to retrieve the CM. At 0831, or just one minute after Admiral McCain departed the ship, Wasson began to steer toward the module from a range of 2,500 yards.

At that time, Captain Seiberlich assumed the conn and carefully monitored the speed of *Hornet* to gradually close on the CM. At his direction, wood chips were thrown into the water from the wing of the navigation bridge to aid in judging the ship's speed through the water.

The approach could be safely made from downwind, since the module had been decontaminated. At 0842, the MK56 GFCS in Radar 51, under the command of Lt. (jg) Chuck Rand, lost contact with the CM at a range of three hundred yards. Smoke flares were ignited to provide a visual wind line for the captain, and a flagman positioned on the forward starboard catwalk signaled to the navigation bridge that the module had passed the bow. The module slowly passed down the starboard side of the ship as *Hornet*'s speed was halted.

At 0844, the shot line was fired toward the outstretched target of Mike Mallory's circled arms. It was a bull's-eye. Within just a few minutes, the in-haul team pulled *Columbia* inboard and directly beneath the B&A crane for the hoisting operation. Lieutenant (jg) Beaulieu was a deck officer in charge of one of the teams which stood ready to hoist and guide the CM onto its dolly, on deck. He recalled that the wind blew at about twenty knots during the retrieval operation, and that waves, of five to six feet, were heavier than expected. The CM was bouncing up and down in the swells and tossed about by the waves, despite the flotation collar. A frogman stood atop the module, on its apex, appearing to ride the module like a bucking bronco, with one hand in the recovery loop and the other reaching to snag the hook at the end of the B&A crane's hoisting cable. That is when someone noticed the daisy—a big bright yellow daisy. It was glued to the middle of John Wolfram's chest. A smaller red daisy adorned his right thigh; and below that, a small orange one. The Navy commando looked like a freshly shorn but quite psychedelic hippie riding a bucking bronco at the rodeo.

Back home in Wisconsin, the *Milwaukee Journal* ran an article about Wolfram which featured the daisies. The article painted the story of a once rebellious, long-haired high school athlete—a hippie-type—who made good by becoming the rugged Navy commando who seemingly retained an element of the hippie within his soul, but who nevertheless was selected by the Navy to participate in the recovery of Apollo 11. It was a good story. The editor had no idea how close to the truth his message ran.

But the hippie's job was not yet done. Lieutenant (jg) Beaulieu recalled that everyone on deck held their breath as Wolfram reached for the hook, dangling at the end of the B&A crane's cable. The rough sea state bounced the module up and down, back and forth—it seemed unlikely that the swimmer would be able to successfully join the hook to the CM's recovery loop for hoisting to begin. He had but once chance to do so. If he missed, then *Columbia* would continue to drift along, down the ship's starboard side. *Hornet* would have to come around again for another try, in a maneuver which would require about an hour. And no one wanted that kind of an ending to an otherwise perfect performance. At the same time, with one little slip—if the swimmer failed to remove his hand quickly enough after slipping that hook over the recovery loop—his hand would be crushed. A passing wave raised him suddenly to within reach of the hook, and as it receded and the CM dropped suddenly into a trough, Wolfram grasped the dangling hook and slipped it effortlessly through the recovery loop at the very last second. Beaulieu stated that Wolfram's performance atop the CM that day represented the finest demonstration of hand-eye coordination he had ever personally witnessed. And after a trip of nearly half a million miles, during which the module employed the most advanced technology then available to accomplish what seemed to be the impossible task of flying to the moon and back, it was finally manhandled in the most low-tech manner by members of the Hornet Module Retrieval Team, lowered onto a dolly, and rolled into the ship's hangar bay.

The time was 0855. Only eleven minutes had elapsed from the time the shot line was fired to the swimmers until the module was placed onto its dolly. Just five months earlier, the crew had spent nearly two and a half hours in a failed attempt to retrieve a similarly configured craft, a floating HS-2 Sea King helicopter, from the surface of the sea. The helicopter had foundered and sunk. The difference between the two sharply contrasting results was summarized most succinctly by Captain Seiberlich exactly three weeks earlier, on 3 July 1969, during the first formal training session for the Hornet Module Retrieval Team, when he projected, "You will retrieve this thing as you train." Preparation and training were keys to success—that, and an expectation of excellence which NASA brought to the table.

Neither the Apex cover to the CM nor the main parachutes were ever recovered. Swim ONE, the UDT-12 team lead by Lt. (jg) John McLachlan, did get its feet wet that day when it recovered the cover for one of the drogue parachutes. Mitch Bucklew was the lucky member of Swim ONE who retrieved the cover from the ocean. It was yeoman's work, and small consolation for a team that had arrived first at the splashdown site, but then missed its chance to demonstrate that it too could perform on the Big day.

Clancy Hatleberg and members of Swim TWO climbed up their cargo net and then a Jacob's ladder onto the ship. A CBS reporter informed Hatleberg that news anchorman Walter Cronkite wished to interview the BIG swimmer during a live broadcast to millions of Americans. But Hatleberg declined the offer. NASA's assistant team leader, John Stonesifer, had directed Hatlberg to speak with Stonesifer as soon as he boarded the ship. Stonesifer needed to ensure that the decontamination process had indeed been completed according to the protocol established by NASA and the ICBC. He queried Hatleberg about the vents, the side crew hatch, and every aspect of the operation. Stonesifer was very detail oriented. When he was satisfied that everything had been accomplished according to plan, he conferred upon Hatleberg: "Good job."

The recording element of Hornet's Module Retrieval Team set a deadline of 0900 for all photographers to turn their film in to a collection point for processing. AN Lee Bigbee and PH3 Tommy Milan spent several hours processing the film.

One member of Hornet's photo lab recalled that a UDT frogman entered the lab to submit for processing a roll of film from an aerial camera. It was Mitch Bucklew, a member of Swim ONE. Bucklew's black-and-white photos were taken before sunrise, in very dimly lit conditions, with a fixed-focus handheld KB-50 aerial camera provided to him by a buddy stationed at the photo lab at the Naval Air Station North Island, near Coronado, California. One of Bucklew's photos captured the very moment when Neil Armstrong, Buzz Aldrin, and Mike Collins returned to earth from humankind's first walk on another celestial body—the splashdown of Apollo 11's command module, Columbia. The photo would have great historical value.

Remarkably, the photo was not among those that were disseminated to the public. In fact, very few people were aware of the existence of the photo until its discovery during research for this book. Mitch Bucklew, who took the shot, did not retain a copy of it. PH3 James Lee Dewey claimed to have seen the photo, but did not obtain a copy either. A duplicate of the photo was mailed to former Hornet photographer's mate, PH2 Douglas Downes, but Downes misplaced it. Downes recalled that the glossy was very small, approximately three inches square, which was consistent with the size of negatives produced by Bucklew's camera. Fortunately, one member of *Hornet's* photo lab retained a copy of the photo, kept it safe for thirty-nine years, and donated the photo for this publication. Bucklew, Dewey, and Downes viewed the photo, some thirty-nine years after the splashdown of Apollo 11's command module, and acknowledged that it does indeed depict the very moment that President John F. Kennedy's goal was finally accomplished—to not only place a man on the moon, but to then bring him safely back to earth.

Members of *Hornet's* photo lab were sitting on a gold mine. During the following two days, while *Hornet* steamed for Pearl Harbor, Hawaii, a flurry of bartering took place on the ship, as everyone scrambled for souvenirs. Members of the press pool held stores of ABC pens, CBS pins, and NBC paraphernalia to trade. The best form of currency in that barter environment, though, was a good quality photo, to which members of the ship's photo lab had ready and plentiful access. They made out like bandits.

Around noon on Splashdown day, ABC reporter Keith W. McBee walked to the XO's stateroom to request a favor. McBee had in his possession the two-inch magnetic reels onto which ABC had recorded the day's events—from the time Marine One landed on *Hornet* at 0512 that morning, to the presidential welcoming ceremony for the astronauts in hangar bay #2, and to the retrieval of the CM just before 0900. Altogether, the film ran to four and one-half hours in duration. ABC had created two original films by routing input from their fixed and mobile cameras to two concurrently operating recorders. The film was then wound into approximate two-hour segments and placed into canisters, each of which weighed about fifty-five pounds. McBee had four of them when he reported to the XO's stateroom. And, he had his personal luggage, as well. He needed help.

McBee planned to catch a COD flight to Pago Pago, and from there he intended to fly directly to Honolulu so that ABC could begin the processing and disseminating of its film. But he had serious concerns about the security of the film. He feared that, if the COD flight went down and crashed into the ocean for some reason, and sank, then irreplaceable film of great historic value would be forever lost. He did not want to take that chance. He asked the XO, Cdr. Chris Lamb, to accept one set of film canisters for safekeeping aboard *Hornet* until the ship reached Pearl Harbor, at which time ABC would receive the canisters if McBee's plane was lost at sea. Lamb agreed and tasked his orderly to store the canisters under the bunk in his stateroom. Lamb's orderly then assisted McBee with the rest of his luggage.

McBee did not call upon Lamb after *Hornet* reached Pearl Harbor. And Lamb was too preoccupied to remember the film tucked under his bunk. A year later, when *Hornet* was decommissioned in Bremerton, Washington, Lamb directed his orderly to pack the personal possessions in his stateroom because Lamb had been transferred. His orderly boxed things up, sealed them, and arranged for shipment.

Twenty-five years later, Chris Lamb received an invitation to attend USS *Hornet*'s twenty-fifth anniversary reunion and celebration of the flight of Apollo 11. He went out to his garage to open some boxes from those days which, he thought, might contain memorabilia that would be of interest at the reunion. That was when he discovered the film.

He questioned its condition and quality after twenty-five years in his garage, so he took it to a commercial firm in Washington, D.C., which pronounced it to be in near-perfect condition. They produced both Beta and VHS versions of the film, then cleaned and conditioned the film for further storage, and finally repackaged and sealed new containers to preserve the film.

In January 2007, Lamb read an article in the *Washington Post* concerning the loss of the videotaped images of the first step on the moon by Neil Armstrong. NASA had stored them but thirty-six years later could not find the tapes. The videos were apparently lost forever. Lamb contacted the NASA Goddard Space Flight Center engineer mentioned in the article (Richard Nafzger) and told him of the Apollo 11 recovery tapes he had preserved. Mr. Nafzger and Mr. Stanley Lebar visited Lamb to view the VHS copy of the original tapes. They agreed that the tapes should be preserved for posterity. They also believed that ABC had long ago lost their original copy of the recovery film. Lamb, it seemed, had the only original film that depicted the triumphant return of Neil Armstrong and his fellow astronauts from the flight of Apollo 11 and the ceremony with President Richard M. Nixon after splashdown.

Lamb was both appalled and dismayed. The flight of Apollo 11 was arguably the most significant technological achievement in the history of humankind, and his small role in the final phase of that flight was certainly one of the most significant events of his own life. Lamb appreciated the value of history. His mother had served as a model for Rosie the Riveter posters during World War II, and he was proud of her role in supporting that war effort. He understood that the film must be safeguarded in a museum for posterity.

He contacted NASA and offered to donate the film to the Goddard museum. But NASA deferred to the Smithsonian Institution as the proper home for a film of such historic value. Smithsonian's attorneys correctly demurred in a case where ownership of the film was questionable. They agreed to accept the film only if Lamb could produce a document in writing which proved his ownership of the film. Unfortunately, Keith W. McBee passed away years earlier.

At the time this author learned of the film's existence, Lamb stored the film canisters under an end table, next to the couch in the living room of his home in Maryland. They have since been relocated to a cool and secure area, but their final status had not yet been established at the writing of this book.

The evening of the recovery, a special splashdown dinner was prepared for the crew. During the stopover in Hawaii, the food services division had purchased sufficient stores to serve shrimp cocktail and fillet mignon for two thousand men. The printed menu included a biography of each of the Apollo 11 astronauts, who took their meals inside the MQF.

And while they ate, the first load of moon rocks was en route to Houston, Texas. A Hornet COD piloted by Cdr. John S. Oster and Lt. Roy Knaub launched from *Hornet* at 1800. On board were several special canisters of moon rocks which had been off-loaded from *Columbia*, packaged and decontaminated for transport by John Hirasaki in the MQF, and carried to the flight deck by young NASA engineer Randy Stone. Two USMC non-commissioned officers accompanied the flight as a security escort. Oster and Knaub flew to Johnston Island. On landing, a USAF officer signed for the cargo. A USAF team loaded the canisters onto a USAF cargo aircraft and flew directly to Houston, Texas, for processing by NASA.

Most of the work was done. *Hornet* had only to return the Apollo 11 astronauts to Hawaii. Commander Stuntz, the officer in charge of the Hornet Module Retrieval Team, recalled, "The entire crew was tired. But everything happened just as the crew had trained. When all was said and done, and the pressure was off, the crew expelled a collective sigh of relief."

Hornet set a course for the two-day trip to Hawaii. The Apollo 11 astronauts busied themselves by signing autographs. Lt. Cdr. Joseph A. Fidd, supply officer, recalled that the astronauts were willing to sign a limited number of envelopes for members of the crew and the civilians on board. The signed envelopes were taken back to Houston with the astronauts until their twenty-one-day quarantine period expired and were then mailed to recipients.

JO2 Chauncey "Chan" Cochran recalled that many men on board *Hornet* availed themselves of the opportunity to walk down to the hangar bay just to catch a glimpse of the astronauts inside their MQF. Cochran was from Ohio, home base of Airstream Inc, which manufactured the specially configured MQFs, so he was both proud and curious enough to pay a visit. MM3 Clifford R. Burr recalled that the astronauts sometimes played around with their visitors by blowing "Blowfish" against the window at passing sailors. Their MQF became a very public bedroom, and after a while the astronauts closed the curtain on their window to secure some privacy.

Sgt. Joe Holt, a member of *Hornet*'s Marine detachment, stated that he often made his rounds on the hanger bay to check on the status of his men, who guarded the CM and MQF twenty-four hours per day until the ship reached Pearl Harbor. Recalled Holt,

> Every so often during a shift I would check up on the sentries, to ensure that all was well, and to determine whether any of them needed to make a head call—stuff like that. The first night after the recovery I came up the ladder to see my sentry leaning against the MQF talking on the phone! (There was a phone on the starboard side of the MQF for communication with the astronauts.) This was incorrect, on so many levels. I nearly had a stroke. As I approached the sentry (I'm pretty certain it was Terry Arbuckle) he immediately straightened up and stood away from the MQF, but he still didn't know what to do with the phone. I grabbed

it from his hand and immediately hung it up, saying something like, "I'm sorry, sir," or some such.

As it turned out the astronaut had wakened and simply wanted to talk to somebody—the others still being asleep in their racks. He'd tapped on the window and motioned for the sentry to pick up the phone. I can well imagine the thoughts that went through my sentry's noggin. What was he supposed to do, ignore the astronaut? Actually in truth he was required to call his Corporal or Sergeant of the Guard (me), but instead he picked up the phone. Whoever was on the other end obviously charmed the hell out of my sentry because he was leaning up against the MQF with a big smile on his face when I first saw him. A big no-no. So I can always tell my grandkids I hung up on the first astronauts to walk on the moon. (Author interview of Joe Holt, 25 July 2009)

QM3 Mike Mallory, a member of Swim TWO, the UDT-11 team which executed the recovery of Apollo 11 after splashdown, recalled walking down to the hangar bay during the return trip to Hawaii. He had walked past on previous occasions to snap photographs of the astronauts inside their MQF, and he hoped to persuade them to mug for him. The Apollo 11 command module, *Columbia*, rested on her dolly a short distance from the MQF which housed the astronauts. *Columbia* and the MQF were connected by a clear plastic tunnel which was used to transfer moon rocks from the CM to the MQF. The tunnel ensured that lunar pathogens would not escape into the surrounding atmosphere within the hangar bay. As Mallory was about to enter hangar bay #2, an emergency door lowered, sealed the entire bay, and thereby denied him entry. He waited a short while. When the door lifted again, Mallory walked over to the MQF. The tunnel had been removed.

Indeed, NASA engineer Randy Stone had stripped the common duct tape that sealed the tunnel to the CM, and then folded it like an accordion back into the MQF for storage. Once the tunnel's sheeting was pushed into the MQF, Stone simply sealed the trailer's outer door. Mike Mallory walked into the area shortly after the emergency blast door was lifted.

As a member of UDT Detachment Apollo, Mallory was steeped in Apollo decontamination procedures. He had spent many months in training for the recovery, was very aware of the concern about back contamination and the precautions that NASA had taken to ensure that lunar pathogens would not contaminate the earth, and had received his own share of reminders from Dr. Stullken and other members of the NASA team during training for violating one or another aspect of the decontamination protocol. Mallory did not know where the tunnel had gone. But its absence gave him an idea.

Mallory found Dr. Stullken seated in the chow hall with several other members of the onboard NASA team. The frogman casually mentioned to Old Baldy that the tunnel between the CM and the MQF had been removed, and that he thought he had seen it discarded in a corner of the hangar bay. That was not true, of course, but Mallory the prankster could not resist the temptation to pull Dr. Stullken's leg. Mallory then asked whether NASA's violation of the decontamination protocol would force the ship to remain at sea throughout the twenty-one-day quarantine period, rather than off-load the astronauts as planned, on their arrival in Pearl Harbor. Dr. Stullken instantly bolted from his chair and made tracks for the hangar bay. Mallory chalked one up for the swimmers.

JO2 Michael R. Wheat was technical director of the ship's closed-circuit television station, TV3, which ran movies and selected television programs provided by the Navy, and broadcast news, sports, and weather for the crew. Wheat recalled that NASA made a kinescope of the astronauts' walk on the moon and flew that film to *Hornet* while the Apollo 11 crew was on their way back to earth. It arrived in the afternoon on Splashdown day. That evening, Wheat ran the film a couple of times over the ship's TV system.

The TV station was connected to the ABC TV pool system and was patched in to a television installed inside the MQF. About 2300 the telephone rang in the TV3 studio and Wheat answered the call. It was Mike Collins, up in the MQF, and he asked Wheat to run the kinescope again. Wheat said, "Yessir!" and spooled it up. Two more times the MQF called, asking to run the film, and Wheat did so. It occurred to Wheat then that the men in the MQF, including Armstrong and Aldrin, the astronauts who walked on the moon, had never seen what most of the world had witnessed, live, four days earlier.

FRIDAY, 25 JULY 1969

Friday was spent by many men on board *Hornet* securing mementos of the trip. The ship was scheduled to arrive at Pearl Harbor in the morning, next day, so little time remained for bartering. One item of particular value was kapton foil from *Columbia*. ETR3 Dennis R. Moran recalled that sailors would double-team the USMC guards posted around the CM. While one or more sailors distracted a guard by engaging him in conversations, other sailors would sneak inside the painted security line to strip small sections of kapton foil from the module.

Of course, USMC guards were in the best position to collect such mementos from the CM. Commander Lamb, the executive officer of *Hornet*, recalled that his orderly happened to be a member of *Hornet*'s MARDET. One morning, his orderly presented to Commander Lamb a gift—not only a section of kapton foil from the module, but a small screw from the side crew hatch. The module was constructed of some three million parts. Guards apparently imagined that a minor disassembly of the CM itself, then, would hardly be noticed. And they were correct. No one from NASA ever called them to account for the missing parts.

Group photos were taken in front of the command module *Columbia*. The men of UDT Detachment Apollo took one such photograph, and even the photographers on board ship did so. After the latter photograph was taken, PHCS Bob Lawson tore a sheet of kapton foil from the module and distributed a small portion to each photographer.

Ens. William C. Whitman, the young officer who reported for duty on board *Hornet* in May shortly before she was designated to serve as the primary recovery ship for Apollo 11, and who was assigned to assist *Hornet*'s public affairs officer throughout the recovery, also served as managing editor for a special USS *Hornet* Apollo 11 recovery mission cruise book, which was prepared by members of the crew to commemorate the accomplishment. Whitman was the last line officer to join *Hornet*'s crew before the ship was decommissioned in June 1970. JO2 Michael Wheat edited the book, of which only a few thousand copies were made and distributed. Most of those books remain in pristine condition, stored in the garages, attics, personal libraries, and basements of former members of USS *Hornet*'s crew.

Captain Seiberlich also kept busy. He took the initiative to mail a package of mementoes to Vice President Spiro Agnew. Lieutenant Dronzek, his administrative assistant, assembled a variety of gifts which included a USS *Hornet* flight deck cap, a First Day cover, and a large inscribed coffee mug. Dronzek mailed the gifts to the office of the vice president of the United States and later received a thank you note from Vice President Agnew's administrative assistant, Stanley Blair. Dronzek observed that, taking the initiative to provide the gifts to the vice president exemplified Captain Seiberlich's habit of paying attention to detail. Dronzek kept the thank you note as a souvenir.

Captain Seiberlich also personally responded to VIP requests for philatelic mail. The Vatican later mailed the following response: "The Secretariat of State is graciously directed by the Holy Father to acknowledge receipt of the special philatelic envelope from the Captain and Crew of the USS *Hornet* . . . and in expressing His sincere appreciation of the loyal filial devotion which prompted this gesture, has the honor to convey, in pledge of abundant divine graces, the paternal Apostolic Benediction of His Holiness." AOCM Norman Charles Woods, who volunteered to process philatelic mail on Splashdown day, recalled that one envelope was processed for former First Lady Jacqueline "Jackie" Kennedy, the wife of the man who set the entire manned spacecraft program in motion.

At sunset that day, the day before *Hornet* pulled into Pearl Harbor to off-load the astronauts within their MQF and to end *Hornet's* role in the recovery of Apollo 11, the ship held an Apollo 11 Thanksgiving service. The COMNAVAIRPAC band, which had provided entertainment throughout the recovery period and supported the welcoming ceremony for the astronauts, played for the service. Captain Seiberlich hosted the event, which was led by ship's chaplains, Lt. Cdr. John A. Piirto and Lt. Francis S. Haryasz, along with their assistants, Seamen Mirsky, Sol, and Russo.

At 0832 on Saturday, 26 July 1969, *Hornet* moored at Pearl Harbor. A banner affixed to her superstructure announced the successful accomplishment of her mission: Hornet + 3.

The astronauts of Apollo 11 were safely home.

Bibliography

U.S. GOVERNMENT PUBLICATIONS

NASA report. Apollo 10 Mission Report, MSC-00126. (http://history.nasa.gov/
alsj/a410/A10_MissionReport.pdf)

NASA report. NASA Apollo Program Summary Report, serial NASA-TM-X-
68725, JSC-09423, dated April 1975. (http://history.nasa.gov/alsj/alsj-
JSC09423.html)

NASA report. NASA/CR-2004-208938, "Lunar Receiving Laboratory Project History," June 2004. (http://www.lpi.u.s.ra.edu/lunar/documents
/lunarReceivingLabCr2004_208938.pdf)

NASA report. NASA/MSC-01856 Apollo Recovery Operational Procedures
Manual, dated 21 June 1971. (Courtesy of Cpt. Chris W. Lamb, USN
(ret.])

NASA report. "The Vision for Space Exploration," February 2004. (http://www.
nasa.gov/pdf/55583main_vision_space_exploration2.pdf)

U.S. Navy document. USS *Hornet* (CVS-12) AIRNOTE 3120, Subj/Loading
Plans and Space Allocation Bill in Support of Apollo 11, dated 24 June
1969. (Courtesy of Cdr. Irwin "George" Patch Jr., USN [ret.])

U.S. Navy document. USS *Hornet* (CVS-12) AIR PLAN for THUR 24/25 JUL
1969. (Courtesy of Cdr. John S. Oster, USN (ret.])

U.S. Navy document. USS *Hornet* (CVS-12), *Hornet* note 1301, Subject: Officer
Duty Assignments, dated 1 October 1969. (Courtesy of Cdr. Harley L.
Stuntz, USN [ret.])

U.S. Navy document. USS *Hornet* (CVS-12) LOGISTIC AIRCRAFT SUP-PORT Form. (Courtesy of Cdr. John S. Oster, USN [ret.])

U.S. Navy document. USS *Hornet* (CVS-12) OPS/ASW Briefing FORM 3840/1 dated 24 July 1969. (Courtesy of Cpt. Chris W. Lamb, USN [ret.])

U.S. Navy document. USS *Hornet* (CVS-12) Splashdown day posting for 24 July 1969. (Courtesy of CW1 Donald W. Harmer, USN [ret.])

U.S. Navy file. USS *Arlington* (AGMR 2) Command History dated 14 January 1970. (Ship's History file, Naval Historical Center, Washington Naval Shipyard, Washington, DC)

U.S. Navy file. USS *Hornet* (CVS-12) Command History. (Ship's History file, Naval Historical Center, Washington Naval Shipyard, Washington, DC)

U.S. Navy radio communication. CINCPAC message dtg 110102Z AUG 69, Subj: Recovery of Apollo 11 and visit of President Nixon. (Courtesy of Cpt. Chris W. Lamb, USN [ret.])

U.S. Navy radio communication. CTF 130 message dtg 232354Z JUL 69, Subj: RETURN OFAPOLLO 11 ASTRONAUTS. (Courtesy of Cpt. Chris W. Lamb, USN [ret.])

U.S. Navy radio communication. CTF 130 message dtg 240358Z JUL 69, Subj: TARGET POINT UPDATE. (Courtesy of Cpt. Chris W. Lamb, USN [ret.])

U.S. Navy radio communication. USS *Hornet* message dtg 200029Z JUL 69, Description: news release by Alexander Volodin, Voice of America. (Courtesy of Cpt. Chris W. Lamb, USN [ret.])

U.S. Navy radio communication. USS *Hornet* message dtg 230712Z JUL 69, Description: news release by Alexander Volodin, Voice of America. (Courtesy of Cpt. Chris W. Lamb, USN [ret.])

U.S. Navy radio communication. USS *Hornet* message dtg 240921Z JUL 69, Description: news release by Alexander Volodin, Voice of America. (Courtesy of Cpt. Chris W. Lamb, USN [ret.])

U.S. Navy radio communication. USS *Hornet* message dtg 231925Z JUL 69, (Section II of II). (Courtesy of Cpt. Chris W. Lamb, USN [ret.])

U.S. Navy radio communication. USS *Hornet* message dtg 231925Z JUL 69, Subj: PRESREL APOLLO RECOVERY WRAPUP. (Courtesy of Cpt. Chris W. Lamb, USN [ret.])

U.S. Navy radio communication. USS *Hornet* message dtg 162104Z JUL 69, Subj: PRESREL—DAILY WRAPUP 15 JUL 69. (Courtesy of Cpt. Chris W. Lamb, USN [ret.])

U.S. Navy radio communication. USS *Hornet* message dtg 170849Z JUL 69, Subj: PRESREL—DAILY WRAPUP 16 JUL 69. (Courtesy of Cpt. Chris W. Lamb, USN [ret.])

U.S. Navy radio communication. USS *Hornet* message dtg 220934Z JUL 69, Subj: PRESREL—DAILY WRAPUP 21 JULY 69. (Courtesy of Cpt. Chris W. Lamb, USN [ret.])

U.S. Navy radio communication. USS *Hornet* message dtg 200943Z JUL 69, Subj: QUARANTINE OPERATIONS MESSAGE NR 2. (Courtesy of Cpt. Chris W. Lamb, USN [ret.])

U.S. Navy radio communication. USS *Hornet* message dtg 221729Z JUL 69, Subj: QUARANTINE OPERATIONS MESSAGE NR 3. (Courtesy of Cpt. Chris W. Lamb, USN [ret.])

U.S. Navy radio communication. USS *Hornet* message dtg 181806Z JUL 69, Subj: SWIMMER, HELO QUARANTINE PROCEDURES. (Courtesy of Cpt. Chris W. Lamb, USN [ret.])

U.S. Navy record. USS *Hornet* (CVS-12) Deck Log, 1969. (National Archives and Records Administration, 8601 Adelphi Road, College Park, MD)

U.S. Navy report. CTF 130, Detailed report of participation in the Apollo 10 Mission. (Courtesy of Cpt. Chris W. Lamb, USN [ret.])

U.S. Navy report. CTF 130 Public Affairs Kit (Project Apollo). (Courtesy of Cpt. Chris W. Lamb, USN [ret.])

U.S. Navy report. Helicopter Anti-Submarine Squadron FOUR, Command History dated 27 February 1970. (Courtesy of Cpt. Charles B. Smiley, USN [ret.])

U.S. Navy report. "UDT-11/Det Apollo memorandum for Apollo 10," undated. (Courtesy of Lt. [jg] Wesley T. Chesser, USN)

U.S. Navy report. USS *Hornet* Apollo 11 Cruise Report. (Courtesy of Cpt. Chris W. Lamb, USN [ret.])

U.S. Navy report. USS *Hornet* Apollo 12 Cruise Report. (Courtesy of Cpt. Chris W. Lamb, USN [ret.])

U.S. Navy report. USS *Hornet* (CVS-12) 1968-69 WESTPAC Cruise Report, dated 27 April 1969. (Courtesy of Cpt. Chris W. Lamb, USN [ret.])

BOOKS

Blair, Don. *Splashdown! NASA and the Navy*. Turner Publishing Company, 2004.

Haldeman, H. R. *The Haldeman Diaries: Inside the Nixon White House*. Berkeley Books, 1994.

Kennedy, Carrie H., and Eric A. Zillmer, eds. *Military Psychology: Clinical and Operational Applications*. The Guilford Press, 2006.

Newport, Curt. *Lost Spacecraft: The Search for Liberty Bell 7*. Apogee Books, 2002.

Schoenberg, Richard D. *The Only Easy Day Was Yesterday*. Naval Institute Press, 2006.

Sullivan, Scott P. *Virtual Apollo: A Pictorial Essay of the Engineering and Construction of the Apollo Command and Service Modules*. Apogee Books, 2002.

Wilford, John Noble. *We Reach the Moon*. Bantam Books, 1969.

Wolfram, John M. *A Call to Arms*. Fontenot, 2002.

PROFESSIONAL ARTICLES, PAPERS, AND PRESENTATIONS

Banks, Louie M., Maj., U.S. Army. "The Office of Strategic Services Psychological Selection Program." Master's thesis, 1995.

Beaulieu, Peter D., Lt. (jg), U.S. Navy. "USS *Hornet* and the recovery of Apollo 11." Text of briefing slide presentation.

Kowal, Dennis M., PhD. "Managing the Unmanageable: Assessing and Controlling the Undercover Operative."

McCormack, Noel. "The Rescue of Apollo 11." http://libertyyes.homestead.com/Hank-Brandli-25.html (accessed March 2008)

Mills, Lisa J., PhD, and Janet D. Held. "Optimizing U.S. Navy SEAL Selection."

U.S. Navy Selection & Classification Office, Arlington, VA, and U.S. Navy Personnel Research Studies & Technology. http://www.internationalmta.org/Documents/2004/2004072P.pdf (accessed July 2007)

Walkover, L. J., R. J. Hart, and E. W. Zosky. JPL Technical Memorandum 33-425. "The Apollo Command Module Side Access Hatch System." http://ntrs.nasa.gov/archive/nasa/casi.ntrs.nasa.gov/19700012142_1970012142.pdf (accessed August 2007)

NEWSPAPERS AND PERIODICAL ARTICLES

McDonald, Gerald W., Cdr., U.S. Navy (ret). "The Recovery of *Aurora.*" *American Heritage Magazine,* April 2001, Vol. 52, Issue 2.

Peck, John. "Apollo 12 Trailer Resurfaces—at a Fish Farm?" *The Huntsville Times,* August 11, 2007.

Townsend, Dallas. "Assignment Downrange" (Cpt. Chris W. Lamb, USN [ret.])

WEB SITES

Check-Six. "Helicopter 66 Recovery Project." http://www.check-six.com/Helo66/h-index.htm (accessed July 2008)

NASA. "The Apollo 11 Flight Journal." http://history.nasa.gov/ap11fj/ (accessed July 2007)

NASA. "Apollo 11 Technical Crew Debriefing July 31st 1969/Landing and Recovery." http://history.nasa.gov/alsj/a11/a11tcdb.html#160 (accessed July 2007)

USS *Arlington* (AGMR 2). "Manned Spacecraft Recovery Force Pacific (TF-130) Meritorious. Unit Commendation." http://www.ussarlington.com/tf130.htm (accessed July 2008)

USS *Goldsborough* (DDG-20). "USS *Goldsborough* DDG-20 1963–1993 History." http://www.ussgoldsborough.com/dnn/History/tabid/180/Default.aspx (accessed July 2008)

USS *Hornet* Museum. "Apollo 11 & 12 Recovery." http://www.uss-hornet.org/history/apollo/ (accessed July 2008)

USS *Hornet* Museum. "Heritage: The Two Hundred Year Legacy of Hornet." http://www.uss-hornet.org/history/heritage/index.shtml (accessed July 2008)

PERSONAL INTERVIEWS

Abbey, George E.; RD2, U.S. Navy; USS *Hornet,* OI Division (Radar and Lookouts)

Barrett, Richard J. "Rich"; Lt., U.S. Navy; HS-4 pilot for Swim TWO

Bateman, Tyler J.; Lt. (jg), U.S. Navy; USS *Hornet,* S-3 Ship's Stores Officer

Beahm, Roger W.; RM2, U.S. Navy; USS *Hornet*, CR Division (Communications)

Beaulieu, Peter D.; Lt. (jg), U.S. Navy; USS *Hornet*, 1st Division Officer (Deck)

Bigbee, Lee; AN, U.S. Navy; USS *Hornet*, OP Division (Photography)

Boisvert, Joseph Louis; Enlisted, U.S. Navy; UDT-11

Boyle, Patrick L. "Red"; Lt. Cdr., U.S. Navy (ret.); USS *Hornet*, V-2 Division Officer (Catapults and Arresting Gear), COD pilot

Brown, Albert L.; CW3, U.S. Navy; USS *Hornet*, GO Division Officer (Aviation Ordnance)

Bucklew, Mitchell L.; ADJ3, U.S. Navy; UDT-12/UDT Det Apollo, member Swim ONE

Carpentier, Dr. William R.; MD, NASA flight surgeon; WFP

Caskey, Robert D.; Cpt., U.S. Marine Corps; USS *Hornet*, Commander, Marine Corps Detachment (MARDET)

Chesser, Wesley T.; Lt. (jg), U.S. Navy; UDT-11/UDT Det Apollo, Team Leader, Swim TWO

Cochran, JO2 Chauncey "Chan"; JO2, U.S. Navy; CINCPACFLT TDY journalist assigned to augment the USS *Hornet* media center

Connor, Ernie E.; Lt., U.S. Navy; USS *Hornet*, OC Division Officer (Air Control) and Assistant CATCC Officer

Dewey, James; PH3, U.S. Navy; USS *Hornet*, OP Division (Photography)

Dronzek, Henry Francis; Lt., U.S. Navy; USS *Hornet*, Commanding Officer's Executive Assistant (EA)

Dunham, Edward M. "Ned"; Lt. Cdr., U.S. Navy; USS *Hornet*, Assistant Navigator

Elliott, Shirley H. "Lee"; Lt., U.S. Navy; USS *Hornet*, OI Division Officer (Radar and Lookouts) and CIC Watch Officer

Esslinger, Michael; historian and author, formerly affiliated with the USS *Hornet* Museum

Fidd, Joseph Adam; Lt. Cdr., U.S. Navy (ret.); USS *Hornet*, Supply Officer

Fish, Robert "Bob"; Trustee, USS *Hornet* Museum

Flynn, Cathal L. "Irish"; R. Adm., U.S. Navy (ret.); former CO, UDT-12

Foss, William C.; FA, U.S. Navy; USS *Hornet*

Gasho, Michael P.; EM3, U.S. Navy; USS *Hornet*, E Division/Flight Deck (Electrical)

Habener, Stephen A. "China Clipper"; Lt., (MSC U.S. Navy); USS *Hornet*, surgeon

Harmer, Donald W.; CW1, U.S. Navy; USS *Hornet*, Ship's Boatswain

Hatleberg, Clarence James "Clancy"; Lt., U.S. Navy; Detachment Commander, UDT Det Apollo, and BIG swimmer

Hirasaki, John Kiyoshi; NASA civilian, Project Engineer for MQFs

Holmes, Thomas G. "Doc"; HM1, U.S. Navy; UDT-11/UDT Det Apollo, alternate (medical)

Holt, Joe; Sgt., U.S. Marine Corps; USS *Hornet*, Marine Detachment (MARDET)

Houston, Willard S. "Sam"; Cpt., USN (ret.), CO, Fleet Weather Center, NAVBASE Pearl Harbor, HI

Klett, William G.; Lt. Cdr., U.S. Navy; USS *Hornet*, CIC

Knaub, Leroy Henry; Lt., U.S. Navy; USS *Hornet*, V-3 Division Officer (Hangar Deck), and COD pilot

Lamb, Chris W.; Cpt., U.S. Navy (ret.); former Executive Officer (XO), USS *Hornet*

Lauck, James S.; Lt. (jg), U.S. Navy; USS *Hornet*, OOD

Laurence, Michael Lee; EM3, U.S. Navy; USS *Hornet*, E Division/Flight Deck (Electrical)

Lawson, Robert L. "Bob"; PHCS, U.S. Navy (ret.); USS *Hornet*, OP Division (Photography)

Lee, Enoust E.; MM1, U.S. Navy (ret.); USS *Hornet*, M Division (Main Engines and Machinery)

Legare, Arthur Edwin, Jr.; PH1, U.S. Navy (ret.); USS *Hornet*, OP Division (Photography)

Lowe, Mickey Neil; BM2, U.S. Navy; USS *Hornet*, 2nd Division (Deck)

Mallory, Michael G.; QM3, U.S. Navy; UDT-11/UDT Det Apollo, member Swim TWO

Mcbee, Lucy H. "Zu"; widow of ABC newsman Keith Wyatt McBee

McLachlan, John D.; Lt. (jg), U.S. Navy; UDT-12/UDT Det Apollo, Team Leader, Swim ONE

Mcnally, John J. "JJ"; Cdr., U.S. Navy (ret.); USS *Hornet*, Operations Officer

Meisenhelder, William M.; Ens., U.S. Navy; USS *Hornet*, OS Division (Sonar)

Mooney, Howard B.; QM1, U.S. Navy (ret.); USS *Hornet*, Navigation

Moran, Dennis R.; ETR3, U.S. Navy; USS *Hornet*, OE Division (Electronics)

Morris, Homer V. "Mo"; CW2, U.S. Navy; USS *Hornet*, Disbursing Officer/ Aviation Stores

Muehlenbach, Terry A.; PH2, U.S. Navy; UDT-12/UDT Det Apollo, member Swim ONE

Oster, John S.; Cdr., U.S. Navy (ret.); USS *Hornet*, CATCC Officer, COD pilot

Ovard, James E.; Lt. (jg), U.S. Navy; USS *Hornet*, Operations Division

Parolini, Roger J.; PH2, U.S. Navy; USS *Hornet*, OP Division (Photography)

Patch, Irwin "George" Jr.; Cdr., U.S. Navy (ret.); USS *Hornet*, Air Officer

Powers, Richard F. III; Lt. (jg), U.S. Navy; USS *Hornet* liaison to POTUS.

Putnam, Thomas M. "Milt"; PHC, U.S. Navy; CINCPACFLT TDY still-photographer

Rand, Charles S. "Chuck"; Lt. (jg), U.S. Navy; USS *Hornet*, Fox Division (Gun Fire Control)

Renner, Ernest A. "Ernie"; Lt. Cdr., U.S. Navy (ret.), HS-2 pilot

Richmond, Donald George; Cdr., U.S. Navy (ret.); HS-4 pilot for Swim ONE

Righter, Gary P.; ABH1, U.S. Navy (ret.); USS *Hornet*, V-1 Division (Flight Deck)

Robnett, Stanley C.; AWHC, U.S. Navy (ret.); HS-4 hoist operator for recovery helicopter

Rorhbach, Robert R.; Lt. (jg), U.S. Navy; UDT-11/UDT Det Apollo, Team Leader, Swim THREE

Seaton, Richard B.; ANH2, U.S. Navy; HS-4 crewman for Swim TWO

Shirer, Glenn S.; CPO, U.S. Navy; UDT-11

Smiley, Charles B. "Chuck"/"Charlie Brown"; Cdr., U.S. Navy (ret.); former CMDR, HS-4

Stone, Brock Randall "Randy"; NASA civilian, MQF technician

Strawn, William Wesley 'Bill'; Lt., U.S. Navy; HS-4 co-pilot for Swim ONE

Stullken, Dr. Donald E.; NASA Landing and Recovery Division; Team Leader, NASA Recovery Team on board USS *Hornet* for Apollo 11 recovery

Stuntz, Harley L.; Cdr., U.S. Navy (ret.); USS *Hornet*, Weapons Officer

Tasker, William G.; Lt. Cdr., U.S. Navy; USS *Hornet*, Meteorologist

Van Hook, William M.; MMFA, U.S. Navy; USS *Hornet*, M Division (Main Engines and Machinery)

Varley, John C.; CW1, U.S. Navy; USS *Hornet*, Personnel Officer (philatelic mail)

Via, Joseph "Joe"; ADJ3, U.S. Navy; UDT-11/UDT Det Apollo, member Swim THREE

Volkmar, James; 1st Lt., U.S. Marine Corps; USS *Hornet*, MARDET

Wasson, William M.; Lt., U.S. Navy; USS *Hornet*, Fox Division Officer (Gun Fire Control), and Recovery OOD

Wheat, Michael R.; JO2, U.S. Navy; USS *Hornet*, TeeVee3 and PAO liaison

Whitaker, Robert; Cdr., U.S. Navy; CO, USS *Davidson* (DTG-1145)

Whitman, William C. "Bill"; Ens, U.S. Navy; USS *Hornet*, Assistant Public Affairs Officer (PAO)

Wiggers, Francis Earl, Jr.; Lt. (jg), U.S. Navy; USS *Hornet*, Executive Services Officer (ESO), and Postal Officer

Windham, Anse E.; AC1, U.S. Navy (ret.); USS *Hornet*, OC Division (Air Control)

Wolfram, John M.; SN, U.S. Navy; UDT-11/UDT Det Apollo, member Swim TWO

Wylie, William T. "Bill"; ACC, U.S. Navy (ret.); USS *Hornet*, OC Division (Air Control)

Zengerle, Robert W.; MM3, U.S. Navy; USS *Hornet*, M Division (Main Engines and Machinery)

Index

Abelson, Robert I. "Bob," 205–6

Agnew, Spiro, 218

Air Force, U.S.: Aerospace Rescue and Recovery Service (ARRS), xi, 153; Aerospace Rescue and Recovery Wing, 65; para-rescuemen (PJs), xi, 8, 64–65, 159; STAR system, 63–64

aircraft: *Hornet*'s transport of squadrons, 2; IFF pods, 23; search and rescue operations, x, xi, 31; splashdown location and location of *Hornet*, 156, 169, 174

Aircraft Deployed Drift Reduction System (ADDRS), 64–65, 159

Airstream, Inc., 51–52, 163, 214

Aldrin, Edwin E. "Buzz," Jr.: aboard Recovery, 197–98; astronaut pen, 127; BIG suits, donning of, 191; BIG training, 83; career of, 16; daisy sticker on CM, 154; decontamination procedures, 193–97; departure from CM, 191–92; *Eagle* guidance system, 123–24; *Eagle* landing, 124–25; *Eagle* liftoff engine, 126–27; First Day covers, 92; Gemini mission, 16; launch abort procedures, 102; lift rafts and rogue wave, 196–97; moon landing and walks, 16, 125–26, 127, 216; MQF, walk to, 199; return to Earth, vii–viii; salary paid to, 16; selection of, 16; splashdown, photograph of, 178, 210–11; status of, 184, 189; welcoming ceremony, 135

animal missions, 42

Apollo 11 mission: anniversary celebration, 212; crew selection, 15–16; events on Earth during flight, 152; exploitation of interest in, 102; flight and splashdown, vii–viii, 131, 138, 145; global interest in, 17, 127–28; historical significance of flight, 7, 20; importance of, vii; launch abort procedures, 85–86, 99, 102; launch of, 53, 85, 99; liftoff schedule for, 49; media coverage of launch, 101, 102; moon walk during, 125–26, 127, 184, 216; navigation system on, 130–31; telegram from Russia, 147; unity of a divided nation through, 13; water landing procedures, viii; weather conditions, 86, 91, 100, 112, 135–38, 144–45, 177. *See also Columbia* command module; *Eagle* lunar module; moon rocks and dust

Apollo 11 novelty items, sale of, 141

Apollo program: Apollo 1 mission, 7; Apollo 6 mission, 79; Apollo 8 mission, 107–8, 142; Apollo 10 mission, 21, 79, 80–82, 88, 153–54; Apollo 12 mission, 79; Apollo 13 mission, 8; Apollo 15 mission, 9; command modules, upside down (Stable 2) orientation of, 10; flotation collar for spacecraft, 43; return and landing procedures, 9–10

Apollo Recovery Operational Procedures Manual (NASA), 22, 36, 45, 97, 186–87

Arbuckle, Terry, 214–15

Arlington, 108, 142, 145–46, 157–58, 179

Armstrong, Neil Alden: aboard Recovery, 197–98; BIG suits, donning of, 191; BIG training, 83; career of, 16; daisy sticker on CM, 154; decontamination procedures, 193–97; departure from AM, 191–92; *Eagle* guidance system, 123–24; *Eagle* landing,

and landing, 16–17; moon landing, reaction
to, 128–30; motto for, 77, 135, 156, 163,
218; name of, 1, 146; navigation of, 131–33,
150–51, 156–57, 161, 167, 168–69; navigation
system on, 29, 30, 132; newspapers for,
89–90, 91–92; off-loading procedures,
practice of, 70–71; Pearl Harbor, departure
from, 85; Pearl Harbor, return to, 66–67;
Pearl Harbor trip, 31, 32; photo lab, 167,
171, 203–4, 210–11; pilots assigned to,
ace status of, 2; pizza order from, 105–6;
position of at splashdown, 179; praise for
from McCain, 207; presidential SIMEX,
107–10; presidential visits to, 146; public
affairs office (PAO) support for media,
75–76; quarantine contingency plan
for, 19–20, 52, 111–12, 182, 192–93; racial
tensions aboard, 69–70; recovery of
CM, vii, 24; recovery of CM, equipment
and supplies for, 22–26, 27–28, 32, 36,
38–39; recovery operation, success of, 102;
resupply of, 70–71; return to Hawaii, 214,
216–17; reunion of, 212; San Diego trip, 29,
30–31; scullery discharge, 98; searchlights,
24–26, 59–60; security concerns, 19;
selection of, vii, 1, 7, 21; Shellback
Ceremony, 99–100; speed increase to reach
splashdown location, 169, 175; speed of,
150–51; splashdown location and location
of, 156–57, 161, 168–69, 173–74; splashdown
location, arrival at, 132–33, 151; splashdown
location, change of, 144–45; splashdown
location, travel to, 106, 130, 185, 188, 189;
staterooms, 72; storage space, 27; supplies
for, 105; talent show on, 141; telegram from
Russia, 147; Thanksgiving service, 218;
Tilley crane, 38, 44, 62–63, 117–18; turnover
of ship, activities during, 3; underway
replenishment operations, 139–40; Vietnam
War operations, 2; whaleboats, 45, 56, 198;
World War II operations, 1–2
Hornet Apollo 11 Cruise Report, 35, 53
Hornet Apollo 11 recovery mission cruise book,
204, 217
Hornet Module Retrieval Team: briefing of,
36, 41; capabilities of team members, 35;
command module retrieval team, 33–34;
communication procedures, 66, 86–87,

104–5; exercises (SIMEX), 65–66, 85,
92–98, 102–4, 115–16, 117–18, 134, 138–39;
improvement of, 60; leadership of, 32, 47;
performance of, 66, 104; practice of, 49;
recording team, 35; recovery and retrieval
procedures, 36–37, 43–44; security team,
34, 36, 46–47; selection of team members,
32, 33–34; tracking and acquisition team,
34; training of, 43–45; training of, at-sea
recovery, 53–63; training of, importance
of, 44; UDT assistance team, 35; walk-
through exercises, 53; weather during
exercises, 65
Houston, Willard S. "Sam," Jr., 137–38, 145

Identification Friend or Foe (IFF) system,
22–23, 55
Interagency Committee on Back
Contamination (ICBC), 14, 49–52, 141,
163, 190

Jones, Clyde W., 161–62, 202
Jones, Donald S., 80, 86, 110, 149–50, 178,
180–83, 197, 198–99

Kennedy, Jacqueline "Jackie," 218
Kennedy, John Fitzgerald, vii, 5, 12, 101, 117
Klett, William G., 161–62, 199, 202
Knapp, Richard I., 33, 86, 112
Knaub, Leroy Henry "Rob," 73, 213
Kranz, Gene, 124

Lamb, Chris W.: Black Bullets, activities of,
69–70; CM souvenirs, 217; film of Apollo
11 recovery, 211–13; on *Hornet* selection, 1;
jacket for Nixon, 166; meetings on progress
of preparations, 35; mother as Rosie the
Riveter model, 213; navigation of *Hornet*,
151, 157, 167, 168–69; navigation skills of,
30, 132, 157; Nixon, amenity requirements
for, 45–47; retirement of, 1; on Seiberlich,
48; SIMEX and weather conditions, 93;
stateroom for Davis, 140; turnover of
ship, activities during, 3; UDT swimmers,
bravery of, 93; White House press corps,
144
Larson, Charles R. "Chuck," 45–47, 161
Lauck, James S. "Steve," 76, 128, 135, 156, 174

About the Author

SCOTT W. CARMICHAEL is currently employed as the senior security and counterintelligence investigator for the Defense Intelligence Agency. He has accumulated a total of approximately thirty years of service with the Department of Defense. He lives with his wife and three children in Crofton, MD.